Inventing the Recording

Currents in
Latin American
& Iberian Music

Inventing the Recording

*The Phonograph and National Culture in
Spain, 1877–1914*

EVA MOREDA RODRÍGUEZ

OXFORD
UNIVERSITY PRESS

Oxford University Press is a department of the University of Oxford. It furthers
the University's objective of excellence in research, scholarship, and education
by publishing worldwide. Oxford is a registered trade mark of Oxford University
Press in the UK and certain other countries.

Published in the United States of America by Oxford University Press
198 Madison Avenue, New York, NY 10016, United States of America.

© Oxford University Press 2021

Library of Congress Control Number: 2021936202
ISBN 978-0-19-755206-3

DOI: 10.1093/oso/9780197552063.001.0001

1 3 5 7 9 8 6 4 2

Printed by Integrated Books International, United States of America

Contents

Acknowledgments

I would like to acknowledge the assistance, encouragement, and advice of the following individuals and organizations:

the Arts and Humanities Research Council, for awarding me a generous research fellowship that allowed me to complete the archival research without which this book would not have been possible;

my colleagues at the University of Glasgow's College of Arts, particularly Dauvit Broun, Martin Cloonan, Björn Heile, and Kirsteen McCue;

the numerous scholars and enthusiasts of early recordings I have had the fortune to discuss my project with over the last few years, including the following: Henri Chamoux, Thomas Henry, Bill Klinger, John Levin, Thomas Y. Levin, Dale Monroe, Bernardo Riego Amezaga, João Silva, Filip Šír, Ulrik Volgsten, Benedetta Zucconi, and, especially, Élodie A. Roy and Inja Stanovic;

Jaione Landaberea at Eresbil, Margarida Ullate i Estanyol (as well as the rest of the staff) at the Biblioteca de Catalunya, the staff of the Biblioteca Nacional de España and of the Museu de la Mùsica;

my parents, brother, and husband.

Introduction

Global technologies, local sounds

Even though Thomas Alva Edison was first able to record and play back sound in 1877, the idea of the music recording as we know it took decades to come into being. Indeed, it was not until the late 1890s that the notion of the phonograph as a domestic machine, which allowed an individual to listen to music in the privacy of their own home, started to impose itself over other competing ontologies of the talking machine, which instead focused on its value as a dictation machine or a scientific curiosity. It was these new conceptualizations of the phonograph that made the very idea of the music recording possible: with phonographs now occupying pride of place inside homes, their owners now demanded a more-or-less constant supply of new recorded music to enjoy. Subsequently, throughout the first decades of the twentieth century and up until the 1930s, a multiplicity of other developments (technological, commercial, social, cultural, musical) further shaped ontologies, practices, and discourses associated with the recording that consolidated its status as a commodity and as a cultural artifact that, while closely connected to live sound, occupied a decidedly separate place in individual and collective consciousness and was bound by a set of generic norms and expectations.[1]

Histories of recorded music tend to start their narratives around 1900—that is, when these new ontologies of recorded sound that tied the phonograph and gramophone to the enjoyment of music and made the idea of the recording possible first emerged.[2] If the period that preceded the era of the record as we know it is covered, it is usually presented briefly and factually, with little critical discussion.[3] One reason for this restrained treatment is that the context in which recording technologies first appeared—scientific demonstrations and magician's spectacles—are difficult to fit with our present understanding of what records and the recording industry are. Moreover, compared to the relatively rich and well-organized archival and business records kept by multinational companies, sources for the study of the earlier period are scarcer and patchier.

Inventing the Recording: The Phonograph and National Culture in Spain, 1877–1914 focuses on the decades in which the recording went from technological possibility to commercial and cultural artifact, and it does so through the analysis of a specific national context: Spain. It tells the stories of institutions and

Inventing the Recording. Eva Moreda Rodríguez, Oxford University Press. © Oxford University Press 2021.
DOI: 10.1093/oso/9780197552063.003.0001

individuals in the country, discusses the development of discourses and ideas in close connection with national concerns and debates, and pays close attention to original recordings from this era. The book starts with the arrival in Spain of notices about Edison's invention of the phonograph in 1877, followed by the first demonstrations (1878–1882) at the hands of scientists and showmen. These demonstrations greatly stimulated the imagination of scientists, journalists, and playwrights, who spent the rest of the 1880s speculating about the phonograph and its potential to revolutionize society once it was properly developed and marketed. The book then moves on to analyze the "traveling phonographs" and *salones fonográficos* of the 1890s and early 1900s; during these years, phonographs were being paraded around Spain and exhibited in group listening sessions in theatres, private homes, and social spaces pertaining to different social classes. It finally covers the development of an indigenous recording industry dominated by the so-called *gabinetes fonográficos*: small businesses which sold imported phonographs, produced their own recordings, and shaped early discourses about commercial phonography and the record as a commodity between 1896 and 1905. Attention is paid as well to the recording multinationals (Gramophone, Odeon, Pathé) who settled in Spain from 1903 onward and forced the *gabinetes*, within a few years, to close down or become resellers of their products.

Readers familiar with the early history of recording technologies elsewhere might have noticed that the preceding narrative includes both familiar and unique elements. For example, "coin-in-the-slot" machines (akin to the Spanish *salones fonográficos*) were commonplace in the United States throughout the 1890s, and they were also the means through which millions of people first became acquainted with the phonograph. Traveling phonographs, on the other hand, were more idiosyncratic to Spain—and, as will be discussed in Chapter 2, their travels were also conceptualized through prevailing understandings of place, geography, and mobility that were unique to the Spanish context. The interplay of these two dimensions (the local-regional-national, on the one hand, and the transnational-global, on the other) is indeed a constant throughout the book. My starting point is not, however, an assumption that Spain's early engagement with recording technologies were exceptional and therefore merit being studied on this account. Instead, I intend to illustrate how the study of a country normally regarded as peripheral, both industrially and musically, but certainly integrated within international networks of different kinds, can shed further light on the plurality of actors, synergies, and discourses that made it possible for recording technologies to be disseminated across the world and absorbed into local practices and discourses. Spanish labels, musicians, and listeners, while being part of a transnational networks and discourses, were also profoundly influenced by their national, regional, and local milieu, and it is this set of complex interactions between contexts that this book intends to make more visible.

The idea that national, regional, and local variables helped shape technological, commercial, cultural, social, musical, performative, and aesthetic aspects in the early history of recording technologies is not new, and it can certainly be regarded as underlying much research on the topic—for example, the multiple histories covering the early history of the recording music industry in particular locations, often written in native languages. However, it is only more recently that scholars have started to explicitly ask themselves how global, transnational, national, regional, and local dimensions interfaced with each other to give rise to practices and discourses that might bear strong family resemblances, but cannot simply be dismissed as local particularities or deviations from the Anglo-American norm. Importantly, such studies, while centering the notion of place, do so from a multiplicity of approaches: for example, Leonor Losa tackles such questions by looking at the early Portuguese music industry from an ethnomusicological perspective,[4] whereas Benedetta Zucconi does so by tracing the early history of the aesthetics of recorded sound in the Italian context.[5] The present book intends to insert some of the same context-sensitive perspectives into a cultural history of early recording technologies in Spain. It does so by using two main questions to drive the historical narrative and facilitate comparisons with other national contexts: Is this development, practice, or discourse specific of Spain, or does it have unique characteristics compared to other countries? If the latter, what are the specificities of Spanish cultural, social, political, or musical history that might explain it?

Although the ways in which the phonograph documented past performance practices constitutes a key pillar of the scholarly study of early recording technologies, this book does not intend to be a study of historical performance practice. Performance practices will certainly make an appearance, especially in Chapters 6 and 7, but what this book intends to be, first and foremost, is a cultural history of early recording technologies in Spain. Performance practice matters will be discussed only inasmuch as they can help us understand such other aspects. I believe that fully understanding the place that recording technologies and recorded sound occupied in the Spanish imagination at this time is fundamental to ascertain the value of these recordings as documents of performance practice, and I therefore hope that this study proves helpful for others intending to pursue such avenues of inquiry in more detail. On the other hand, while documenting performance was clearly a key theme in the early history of recording technologies, we must bear in mind that, in the context of the present study, it is not the only one, and in fact it was not even the main one in some periods of time.

Early recording technologies in Spain: Cultural history, musicology, sound studies

As with the previously cited histories of the recording industry elsewhere in the world, the early history of recording technologies in Spain has not been extensively studied, and most of the efforts in this regard have come from outside the academic research community. Indeed, much of what we currently know about the early history of the phonograph in Spain we owe to collectors and enthusiasts who have uncovered and sometimes contextualized original recordings and factual information, even though these might not always be systematically analyzed and theorized. Key sources in this respect include the newsletter *Girant a 78rpm*, with 17 issues published between 2002 and 2010 by the Catalonia-based Associació per a la Salvaguarda del Patrimoni Enregistrat (Association for the Protection of Recorded Heritage) and numerous short accounts and articles about the early years of recorded sound in Spain, which will be duly cited in the course of this book. Mariano Gómez-Montejano's self-published book *El fonógrafo en España. Cilindros españoles*[6] was the first to present a picture of the *gabinetes*, which some collectors had previously regarded as amateur, non-commercial ventures.[7] The website of collector Carlos Martín Ballester (www.carlosmb.com), dedicated to the selling of early and rare recordings, occasionally features early Spanish cylinders for sale, on which some contextual information is given.

Spanish libraries and archives have also made a crucial contribution to publicizing and disseminating the early history of recording technologies in Spain. In the past decade, several sound collections have been catalogued and, in some cases, digitized. These include, most notably, Eresbil—Archivo Vasco de la Música, which is home to more than 500 cylinders; the Biblioteca de Catalunya, which counts 385; the Biblioteca Nacional de España, with more than 300; Diputación de Huesca, with about 200; and the Museum Vicente Miralles Segarra at the Universitat Politècnica de Valencia, which hosts 61. All of this results in a body of almost 1,500 sound cylinders, the overwhelming majority of which were produced in Spain and therefore constitute a crucial source for this book and particularly for Chapters 3–7. Curators, archivists, and collectors have also authored most of the documentation (websites, CD liners) that has accompanied some of these digitizations, including a range of invaluable contextual information.[8]

As invaluable as the work of collectors, librarians, and curators has been in preserving and uncovering early recordings and providing the first layer of contextualization, what is still overwhelmingly missing is a full critical engagement with the material they have uncovered and with other materials that remain unexplored. We might expect such critical engagement to come primarily from the disciplines of Hispanic cultural studies and musicology; both, however, have some particularities that partly explain why this has not been the case so far. Firstly, the field of Hispanic cultural studies has traditionally focused on the textual and the visual, with sound only very recently starting to attract the attention of scholars. Importantly, some of the first studies to focus on sound that are now starting to emerge cover the same period as this book does (the late nineteenth and early twentieth century),[9] confirming that this was indeed a period where new ways of thinking about and contextualizing sound were quickly developing in close connection to new discourses about science, technology, social organization, and national identity. The early development of recording technologies in Spain should therefore be viewed against this backdrop that is increasingly coming to the attention of researchers.

The field of musicology in Spain, on the other hand, has historically privileged approaches centered around composers and their works, with studies of discography typically being regarded as the province of collectors or music critics. The field, however, has experienced a recent surge of interest in performance studies,[10] and the first studies of early recordings as documents of performance practice have started appearing too.[11] Not all of the early history of recording technologies in Spain falls, strictly speaking, within the realm of musicology: as indicated previously, in the initial 15 years of the phonograph, the idea that it should be used primarily to record music had not fully developed yet, and other types of non-musical sounds were frequently recorded. However, going back to the prehistory of the music recording, as I do here, can inform our understanding of the early encounters of Spanish composers and performers with the phonograph and gramophone throughout the early decades of the twentieth century.

Whereas Spain-specific sources are scarce, there is no shortage of research on the early history of recording technologies in locations other than Spain that this book draws upon. For the purposes of clarity, and to account for the variety of approaches that the complexity of the subject has invited and that are reflected— to different extents—in the subsequent chapters of this book, I divide this research into four broad categories. The first category is concerned with the study of the material, technological, and financial realities of early recorded sound— from the development and spread of technological innovations to the business strategies and economic models adopted by the nascent recording industry. These were indeed key questions for some of the earliest research into the phonograph as it unraveled during the 1960s and 1970s, often conducted by historians and technology enthusiasts, and they have continued attracting attention up to the present in a number of academic disciplines.[12] Although this book is not conceived as solely or even primarily a history of the early Spanish recording industry, it draws extensively on examples and models set out by the previously cited research when engaging with the commercial practices and strategies put in place by the *gabinetes* to turn a curious technological invention into a source of profit and the technological challenges and opportunities they faced in doing so, as well as with the influence that recording technologies had on the working lives of the individuals who worked with them, broadly understood.

The second area of research concerns the study of early recordings as documents of performance practice by musicologists and performance scholars—an area of research that has thrived since the early 1990s.[13] As has been advanced earlier in the Introduction, performance practice is not relevant to all of the issues covered in this book; particularly, it is only from the late 1890s onward that we start to have recordings that are useful in terms of drawing conclusions about historical performing styles. However, these studies become more relevant in the latter part of the book because, in problematizing the role of recordings as performance practice documents, they have contested the notion that historical recordings should simply be seen as a photograph of music-making practices on stage; instead, they have drawn attention to the mechanisms by which recorded music is constituted as its own genre while still being closely connected to live music on stage. By examining Spanish-specific evidence under the light of these studies, in Chapter 6 I offer some guidelines as to how the historical and cultural study of the early recording industry in Spain can inform future studies of performance practice in the country—such as the types of repertoires that were recorded once the *gabinetes* started operation and how commercial recordings fitted into the working lives of singers.

A third area of research concerns the impact of recording technologies on less tangible, more abstract domains, namely on how individuals and societies listened to music, and how they conceptualized both recorded sound, and sound more generally.[14] Such shifts can be seen through the whole period under study, but are more obvious in the *gabinetes* era as the notion of the music record finally crystallized. Finally, a body of research coming mostly from cultural studies and the history of science and technology (rather than from musicology or performance studies) has explored how recorded sound (as opposed to recorded *music*) changed ontologies of sound and introduced transformations in cultural and memory practices;[15] I draw upon this bibliography mainly in the earlier chapters of the book, which cover the era preceding widespread use of the phonograph as a music-playing device.

Setting the scene: *Regeneracionismo* in Spain and the early years of recording technologies

Understanding how local, regional, national, transnational, and global dimensions interfaced in shaping the early history of recording technologies in Spain requires some scene-setting, particularly for readers unfamiliar with Spanish history, the history of early recordings, or both. The accounts I present in this section are to be understood in this spirit, and some of the points I make here will be expanded upon in the following chapters. Regarding Spanish history, the period covered in this book (1877–1914) roughly overlaps with the first two thirds of the so-called Bourbon Restoration (1874–1931)—an era of relatively political stability after the troubles that had plagued Spain from the early nineteenth century onward. In September 1868, Queen Isabella II was overthrown by the Glorious Revolution after a long reign marked by intrigues, political inefficiency, and continuous *pronunciamientos* (military rebellions). The Revolution, however, initially failed to solve Spain's problems: a foreign king, Amadeo I of Savoy, was brought in and reigned for less than two years before he abdicated in February 1873. The ensuing First Republic was similarly short-lived, overthrown in December 1874 in a coup led by General Martínez Campos. Monarchy was then restored in the person of Alfonso XII, Isabella's son, and Spain became a constitutional monarchy, albeit one marked by agreed electoral fraud. Indeed, in order to guarantee political stability, the two main parties (Conservative and Liberal) agreed to alternate in the government for a certain number of years each, in a system known as *el turno pacífico* (the peaceful shift). This required election results to be manipulated—which the parties did mostly through local leaders in the provinces called *caciques*, who, in return for political favor, could enable electoral fraud.

The *turno pacífico* came to an end in the 1914 general election—when, for the first time, neither the Conservatives nor the Liberals were able to obtain a majority. Until then, though, the system had been relatively effective in terms of ensuring political stability: with the liberals being allowed to partake in the government, this diminished the risk of *pronunciamientos* and allowed the government to pursue reform programs and economic development relatively unencumbered. The system, however, also encouraged tensions and resentment to develop among sectors of the population. These included the nascent nationalist movements in Catalonia, the Basque Country, and, to a lesser extent, Galicia, which advocated for increased autonomy and devolution; and also from the increasing masses of workers organized under socialist or anarchist principles.

Another important source of discontent with the *turno pacífico* system—and also the most relevant for the developments discussed in this book—was the loosely organized movement known as *regeneracionismo* ("regenerationism"). *Regeneracionismo* can be broadly defined as the idea that Spain was in crisis and that solutions of some sort needed to be implemented at an economic, political, cultural, scientific, and even existential level.[16] The first proponent of *regeneracionismo* was Joaquín Costa (1846–1911), a lawyer by training, who for years campaigned among politicians advocating for an ambitious reform program pursuing the "de-Africanization and Europeanization of Spain."[17] Costa's program was wide-ranging, encompassing a complete reform of the education system along European lines (under the influence of German philosopher Karl Christian Friedrich Krause), land-owning reform and support for farmers, the development of international trade networks, the stabilization of the monetary system, and increased self-government for municipalities.[18]

While Costa did not start a movement, lobby, or party in the traditional sense, throughout his life his *regeneracionista* focus percolated through Spanish society, and groups and individuals in different social classes, political ideologies, and institutional allegiances started proposing their own programs for national *regeneración*, often working within their own area of interest or expertise. Consequently, the areas that different individuals or groups singled out for *regeneración*, as well as the solutions they proposed, were often very different among themselves and also from Costa's original program.[19] For example, journalist, teacher, and activist Jesús Pando y Valle published in 1897 *Regeneración económica. Croquis de un libro para el pueblo*,[20] where he argued that, even though multiple political reforms had been suggested in the previous years, it was now the time for national economy to be regenerated by having the state more wholeheartedly support the industry (a sector which was absent from Costa's program) and by introducing the scientific and academic study of the economy into Spanish universities. In his 1899 tract *La regeneración. El problema político*, journalist and politician Antonio Royo Villanova admitted that the question of national regeneration was a multifarious one, involving political, economic, demographic, moral, religious, and educational dimensions.[21] Importantly, Costa and other *regeneracionistas* saw themselves as a new, vigorous strand of Spanish society that opposed the Restoration establishment—the fact notwithstanding that some *regeneracionistas*, including Costa, held significant positions in the same establishment they claimed to oppose;[22] the same contradictory attitude toward state institutions can be observed among some *gabinete* owners, as will be discussed in Chapter 3. *Regeneracionismo* became reinvigorated after Spain lost its last colonies in 1898, sending the country into an existential crisis just as other European nations were fully engaged in their colonial enterprises.[23]

Although, as has been noted previously, not all *regeneracionistas* agreed on the solutions that should be adopted to tackle the Spanish crisis, many saw Spain's alleged backwardness in science and technology as one of the reasons which had led the country to lose the Cuban War against the United States, and one of the issues that should be solved as a matter of urgency if national regeneration was to be achieved.[24] As with other points in the *regeneracionista* program, impact on the establishment was mixed: some measures were introduced, especially in terms of reorganizing education institutions to promote the teaching of science in a more practical way, but its effects were not long-lasting or truly transformative.[25]

Regeneracionismo, however, helped put science in the Spanish imagination during the period under study here, and the following chapters discuss how some of their prevailing discourses about science and technology, and their role in modernizing Spain, shaped the early reception of recording technologies. The interest in science, though, did not solely develop from the *regeneracionista* circles: already in the first half of the nineteenth century numerous individuals and institutions throughout Spain strived—often with little institutional support—to disseminate scientific discoveries from abroad, even though considerably less attention was paid to conducting original research.[26] The second half of the century also saw—as will be discussed in Chapters 1 and 2—a surge of popularity for scientific spectacles, mostly of the visual kind. Such spectacles did not pursue reform or national improvement per se, like *regeneracionismo* did, but they too contributed to turn science into a preoccupation for Spaniards. In this context, as will be explored throughout the book, the phonograph was initially saluted as a scientific and technological invention (with little consideration being given, at least initially, to the musical content of the recordings); as such, phonographs were fitted within, and in turn influenced, discourses concerning science and technology and what these could do for national modernization and regeneration. At the same time, some, but not all, of those involved in the early phonograph industry adhered to *regeneracionista* ideals.

Another salient theme in *regeneracionismo*—and in the broader Restoration scene too—was the development of new understandings around place and space, which also impacted significantly on the reception of recording technologies in Spain. *Regeneracionistas* were concerned with mapping out and exploring the physicality of Spain and its nature,[27] overcoming those aspects of its landscape that were seen as an obstacle for development,[28] and allowing other Spaniards to get to know and enjoy their environment.[29] *Regeneracionismo* was also concerned with place and space in Spain in a more abstract way, as the *Desastre* of 1898 and successive military campaigns in Morocco, as well as peripheral nationalisms closer to home, threatened to challenge notions of what counted as Spanish territory.

Two aspects of geography and place, in particular, are relevant to the early reception of recording technologies in Spain. The first of those, which will be most relevant in Chapters 1 and 2, is the notion of the phonograph as a mobile object, with devices paraded around Spain and shown to audiences in itinerant demonstrations. These traveling phonographs indeed provide a fruitful ground for the application of perspectives coming from the developing field of music and mobility, where mobility is not simply as synonymous with movement; instead, it encompass too the fluctuating meanings that musics and musical materials acquire by virtue of being mobile, the discourses connected to movement itself, and the representation of mobility in music.[30] Mobilized elements, such as traveling phonographs in Spain, must therefore be understood as more than vehicles and containers for content: instead, they played an important role in shaping and defining musical and sonic experience.[31]

As will be discussed primarily in Chapter 2, the meanings the traveling phonographs acquired in Spain and the discourses they inserted themselves into were directly connected to *regeneracionista* views of mobility, place, and national identity. Mobility, mostly under the form of transport, was certainly important here: ministers influenced by *regeneracionista* ideas, such as Rafael Gasset (active 1900–1903), focused their attention on developing Spain's transport networks, in the understanding that these measures would facilitate the modernization of Spain through allowing the circulation of goods and people, not only between Spain and foreign countries, but also within Spain. As some *regeneracionistas* noted, the precariousness of Spain's internal communications meant that it was cheaper for some Spanish regions to import wheat from Russia and America than from Castile.[32]

The *regeneracionista* concern with introducing improvements to the transport network and increase mobility as a way to bring over modernization to Spain was not, however, completely new. Investment in railways and roads had started during the reign of Isabella II back in the 1840s, and the developing transport network was quickly used to emphasize ideas of national unity and support for the monarchy. It was railways that acquired the most notable symbolic meanings here: the queen indeed went on a series of train journeys throughout the country's territory, allegedly so that she would get to know her subjects better, and vice-versa; these same ideas of mutual understanding (not necessarily between a monarch and her subjects, but between communities from different parts of Spain) subsequently reappeared in the *regeneracionista* imagination.[33] Through Isabella's travels, and through further initiatives too throughout the second half of the century, trains became a powerful symbol of modernization because they were mobile in multiple ways. Building a locomotive or setting up a railway network, in the first place, invited and encouraged mobility, with Spanish businessmen and government officers visiting international fairs where they could learn about new technological developments and acquire colonial resources.[34] Once built and set up, the railway's role in facilitating mobility and therefore enabling the modernization of the country was often fraught with contradiction. For example, railway building was initially celebrated in the understanding that trains would make Spaniards (and the goods they produced) more mobile; this would contribute to trade and economic development and would also improve relations between the different regions of the country, therefore putting an end to disputes and hostilities.[35] Over time, though, it became increasingly clear that the railway needed a constant flow of investments from abroad, which paradoxically made the Spanish economy less self-sufficient and more dependent on foreign trade.[36] Another area where discourses of mobility and modernization built around trains were ambiguous and contradictory concerned the landscape. In some *regeneracionista* imaginations, Spain's landscape was seen as immutable and eternal, as well as one of the utmost expressions of national identity; in this context, train travel was seen as a positive force because it would make it possible for Spaniards to get to know their own landscape in a quicker and more effective way than simply walking through it—and from a unique perspective too, as more bridges and tunnels were built. At the same time, though, Spaniards were aware that the constant building of railways was irreversibly modifying this allegedly eternal landscape.[37] As will be discussed in Chapter 1 and especially Chapter 2, traveling phonographs also invited complex discourses around their mobile nature and their ability to travel throughout Spain, not only literally, but also metaphorically.

The second area within *regeneracionista* understandings of geography and place that is relevant for the study of recording technologies is urban development, which is introduced toward the end of Chapter 2 (as itinerant demonstrations were replaced by stationary *salones fonográficos*) and becomes more prominent in Chapters 3–5, as the *gabinetes fonográficos* established themselves overwhelmingly in the main cities of Madrid, Barcelona, and Valencia (while itinerant demonstrations continued to an extent in the provinces). As is the case elsewhere in Europe, Spanish cities saw considerable development in the second half of the nineteenth century, which instigated animated debates—mostly at the local level—around the nascent discipline of urbanism. In Madrid, Barcelona, and, to a lesser extent, Valencia, urban development decidedly shaped the working lives of the *gabinetes*. Moreover, the newly developing, fast-growing cities also came to acquire certain meanings—as again will be discussed in due course—that speak of the profound debates taking place in Spain at this time about the very physicality and identity of the country. Madrid was, of course, the capital city of an admittedly centralist state; as such, it was the preferred destination of a multiplicity of ambitious immigrants of all social classes from all over Spain that contributed to shape the city's geography and personality—but it was also a frequent target for the resentment from those regions who aspired at increased self-government. Barcelona, as the capital of one such region, was at the same time establishing itself as a potential competitor to Madrid—not only within Spain, but also internationally. Such rivalries were partially mirrored in the realm of recording technologies—with Madrid leading the way during the *gabinetes* era, but the industry then gravitating toward Barcelona as multinational companies chose to open their offices there.

The particularities of Spanish history and culture, and especially the discourses disseminated by *regeneracionismo*, certainly shaped the early reception of recording technologies in crucial ways, sometimes resulting in practices that were rather unique in the global context. This, as will be discussed in the course of this book, was also facilitated by the fact that Spain, not initially being a prized market for Edison or Berliner in the same way as the United States or the United Kingdom were, was able to develop relatively independently from these multinational companies. Of course, it was not completely cut off either, and indeed technological innovations imported from abroad determined to a great extent how Spaniards interacted with phonographs at any given point. An over-view of such technological innovations is therefore provided here as a backdrop to the events and practices discussed in the subsequent chapters.

American inventor Thomas Alva Edison is generally credited with inventing the phonograph in late 1877, with *Scientific American* announcing in its December issue that "Mr. Thomas A. Edison recently came into this office, placed a little machine on our desk, turned a crank, and the machine inquired as to our health, asked how we liked the phonograph, informed us that it was very well, and bid us a cordial good night."[38] Edison was certainly not alone in his preoc-cupation with mechanically capturing and reproducing sound: indeed, this had long been a concern for many scientists and inventors worldwide, with French amateur scientist Charles Cros describing, in a paper submitted to the French Academy of Sciences, a method to build a similar machine only a few months before Edison manufactured his.

Edison's first phonograph recorded on a thin sheet of metal, normally tin-foil; sound would be projected into a diaphragm connected to a stylus that made indentations into the metal. Tinfoil, however, would rapidly deteriorate after just a few uses. This, and the generally limited quality of the sound, meant that the phonograph was at this time exhibited and demonstrated mostly as a curiosity. It was still not fit to become an office machine for dictation, correspondence, and other purposes, as Edison intended at this time;[39] the idea of using phonographs for recording and playing back music had yet not entered Edison's or most of his contemporaries' consciousness.

For several years, Edison showed little interest in developing his invention further. It was competition that eventually led him to do so: in 1886, Alexander Graham Bell launched an improved machine based on Edison's model which he called the graphophone, and in 1887 German-American inventor Emile Berliner invented yet another sound reproduction device he called the gramophone—which recorded on a vertical-cut flat disc rather than a piece of metal. As a re-sponse, Edison founded the Edison Phonograph Company in October 1887, and successively developed the New, Improved and Perfected Phonograph, which re-corded sound on wax cylinders instead of tinfoil plaques.

The Edison Phonograph Company initially developed the new device in a decidedly recreational direction, with "coin-in-the-slot" phonographs being installed across America, and demonstrations continuing there and extending to elsewhere in the world, sometimes at the hands of Edison's agents, sometimes organized by enthusiastic amateurs.[40] Domestic phonography, on the other hand, did not start being a serious consideration until the mid-1890s, after Edison introduced the Spring Motor Phonograph and founded Edison Records. Simultaneously, others were trying their hand as well at developing the phonograph not merely as a scientific curiosity but as a device facilitating the consumption of music in the home, with the best known of these after Edison being Italian-American engineer Gianni Bettini. Individuals and companies in other countries were doing the same, such as Pathé Frères in France. Concurrently, Berliner had been developing his gramophone for the same purposes, founding Berliner Gramophone in the United States in 1893 and The Gramophone Company Limited in the United Kingdom in 1898 under the direction of Fred Gaisberg.

Unlike the phonograph, the gramophone did not offer the option for customers to make their own recordings. Gramophone discs, however, were cheaper to produce and sell, and so the gramophone ended up imposing itself on the phonograph in most of the world throughout the first decades of the twentieth century—including Spain, where the dissemination of the gramophone brought over the demise of the *gabinetes* industry as early in 1905. While it was the phonograph that initially introduced the idea of recording technologies being used for the aesthetic enjoyment of music and not as a scientific curiosity, it was the gramophone which really made recording technologies commonplace and widespread, and an integral part of the everyday experience for the men and women of the early twentieth century. By 1910, the recording industry had established itself in practically all countries around the globe.[41]

Outline of chapters

While the book is roughly organized in a chronological manner, there are overlaps and duplications between chapters, which should be regarded as a reflection of the complex nature of the early history of recording technologies. Indeed, this history is not one where one invention is replaced by another and then another, but instead one in which practices, institutions, and discourses often coexisted, giving sometimes rise to frictions and competition. Chapter 1 covers the arrival of recording technologies in Spain from the first notices about Edison's phonograph in newspapers and magazines (1877) to the launch of the Perfected Phonograph (1888). It introduces the individuals who first brought the phonograph to Spaniards between 1878 and 1882 through the mean of public demonstrations (Catalan physicist Tomás Dalmau, and showmen Bargeon de Viverols and Dr. Llops), and places them the context of both discourses about science and the politics of audio-visual spectacles in late nineteenth-century Spain. With such public demonstrations practically disappearing after 1882, the chapter then moves on to consider how, nevertheless, recording technologies retained a place in the Spanish imagination, with journalists, cultural commentators, scientists, and even playwrights imagining uses for the phonograph that spoke of their concern with representation—a key component of contemporary discourses about science and modernity that influenced, and were in turn influenced by, the reception of the phonograph in Spain. Chapter 2 covers the developments following the introduction in Spain of Edison's Perfected Phonograph (1888), which opened the door for numerous funfair operators and science enthusiasts to acquire phonographs and parade them around Spain, demonstrating them in theaters and other public venues and organizing private listening sessions for a fee. The chapter discusses three aspects of these demonstrations that connected recording technologies to broader social concerns relating to modernity. Firstly, I discuss how recording technologies expanded and challenged predominant discourses about mobility. Secondly, by looking at phonograph demonstrations in private clubs and societies across all social classes, I analyze how recording technologies shaped and were in turn shaped by discourses concerning sociability and civility. Thirdly, I expand the discussion of representation started in the previous chapter by discussing the listening practices and documented audience responses at phonographic demonstrations.

The book then moves on to consider the beginnings of commercial phonography in Spain, dominated by the *gabinetes fonográficos* between 1896 and 1905. After Edison introduced the Spring Motor phonograph, Home phonograph, and Standard phonograph in rapid succession between 1896 and 1898, the device became relatively easy to operate and affordable to the middle and upper classes. This finally facilitated the development of the record as a commodity: indeed, it was at this time that the appeal of the phonograph shifted from reassuring the audiences of phonographic sessions that it could reproduce sound in an accurate way to providing an aesthetically satisfactory listening experience in the home. Chapters 3–5 discuss the *gabinetes* themselves, and, on account of the significant differences between the operations of the *gabinetes* in different parts of Spain, each chapter presents a different geographical focus, with Chapter 3 addressing *gabinetes* in Madrid. The chapter uncovers a wealth of information about the fourteen studios known to have operated in Madrid, concerning their dates of operation, individuals involved, recording practices, and types of recordings produced. It then discusses the discourses that some of the *gabinete* owners built, through their catalogues, the magazine *El cardo* and other interventions in public life, with the aim of acquiring visibility, marketing their own products, and establishing their identity as innovative, patriotic entrepreneurs in a time in which Spain was fully engaged in discourses concerning national modernization and regeneration following the loss of its last overseas colonies in 1898. The chapter also briefly discusses the disappearance or transformation of *gabinetes* after multinational companies opened subsidiaries in Spain from 1903 onward.

Chapter 4 focuses on the nine *gabinetes fonográficos* active in Barcelona at different points between 1899 and 1904. A significant part of the chapter focuses on the question why the Barcelona *gabinetes* were not as commercially successful as those in Madrid and failed to articulate a discourse that spoke to current concerns about national identity and modernization. The article ends with a brief consideration of how recording activities in Barcelona expanded at the hands of multinational companies (Gramophone, Odeon, Pathé) from 1903 onward, thus temporarily broadening the chronology of the book into the 1910s, during which a second generation of shop owners and entrepreneurs played a crucial role in integrating recording technologies within the morphology and identity of the city. Finally, Chapter 5 moves to the *gabinetes fonográficos* outside the two main cities. It starts by discussing some twenty *gabinetes* known to have been active in provincial capitals and large towns throughout Spain, disseminating in their milieu versions of the discourses concerning science, modernity, and national identity shaped by the Madrid *gabinetes*. Significant attention will be paid to the city of Valencia: indeed, even though it hosted comparatively few *gabinetes* (four), these were among the most active in Spain overall, and, similarly to the Madrid *gabinetes*, they too articulated a discourse mostly through their catalogues and the magazine *Boletín fonográfico*. I will discuss how this discourse, even though similar to the one originating in Madrid in its attempt to connect recording technologies to broader notions of modernity and national identity, had some unique characteristics.

Chapter 6, while staying within the era of the *gabinetes*, focuses on the singers who are known to have recorded for them and helped shape the development of the recording as a commodity in Spain. In these early years, it was not necessarily the best known or most acclaimed *zarzuela* and opera singers who agreed to be recorded, but instead the *gabinetes* relied on a multiplicity of working singers of different standards. A significant part of the chapter will therefore be concerned with reconstructing some of these singers' unknown trajectories and analyzing how they fitted recording sessions within existing labor and musical practices, arguing that those early recordings did not effectively function as a substitute for live performance, but in fact complemented it as a sort of postcard or memento. Drawing on my conclusions about the cultural meanings of recorded music at the time, I will also discuss the value that these early recordings might have in terms of studying and reconstructing performance practice, particularly of *zarzuela*.

Chapter 7 puts the focus on the individuals who bought and collected recordings in Spain in the era of the *gabinetes* and the first years of recording multinationals. Focusing on written sources, as well as on the five surviving re-cording collections whose original owners are known to us, I will discuss what we know about how the first phonograph owners in Spain bought and consumed recordings, how they started to develop new listening habits and practices that would then become widespread, and how they partook in and shaped a culture of home recording that the phonographic magazines encouraged.

Finally, my conclusion will provide further reflection on how my discus-sion of Spain might shed methodological insights as to how to make sense of the differences between the different national contexts, shedding further light as to what was global/transnational in the early development of recording tech-nologies, and what developed differently in different countries, as well as some reasons why this might be the case.

1

Imagining the phonograph, 1877–1888

From the first notices about Edison's *teléfono de repetición*[1] in the industrial magazine *La gaceta industrial*[2] in 1877 to the invention of the Perfected Phonograph in 1888, the very early history of recording technologies in Spain is a history of the fragmentary, ephemeral, and anecdotal: Edison's original phonograph was still too difficult to operate, its playback capabilities too limited, to be developed and widely accepted as a domestic appliance for the regular consumption of aesthetically pleasing auditory experiences. As was the case elsewhere,[3] in Spain Edison's invention was saluted with interest and high expectations, and demonstrations of the phonograph—combining scientific and recreational purposes in different measures—were organized from 1878. From 1882, such demonstrations became less frequent, and serious interest in firsthand engagement with recording technologies did not revive until the launch of the Perfected Phonograph, with demonstrations of touring phonographs becoming a common occurrence in Spain throughout the 1890s.

However, even as opportunities to engage with the device firsthand declined, the phonograph was still being discussed in the Spanish press and scientific circles and reimagined in popular culture. The reason for this apparent paradox has to be sought in the precariousness of the technology (which had similarly discouraged others elsewhere[4]), but also in its promise: Spaniards might not have been particularly interested in listening to the primitive phonograph once they had convinced themselves of its capabilities, but they were still curious and keen to discuss the opportunities and possibilities that further technological developments could bring about in the future.

We can therefore assume that, between 1877 and 1888, only a minimal part of the Spanish population experienced the playback capabilities of the phonograph firsthand, and even fewer would have been exposed to phonographs with some regularity. Nevertheless, the years when sound recording technologies first came into the consciousness of Spaniards deserve attention for a number of reasons. The initial interactions of Spaniards with phonographs—normally under the guise of public demonstrations—indeed confronted them with questions regarding the place of the new invention in contemporary discourses about technology and modernization, its potential for entertainment, its influence in redefining listening practices and notions of representation, and its role in questioning the boundaries between public and private spaces.

Inventing the Recording. Eva Moreda Rodríguez, Oxford University Press. © Oxford University Press 2021.
DOI: 10.1093/oso/9780197552063.003.0002

Moreover, some of these questions, debates, and imagined uses of recording technologies found continuity in the following decades, first as phonographic sessions became commonplace during the 1890s and then as the phonograph and gramophone established themselves as domestic technologies and a staple of consumer culture throughout the first decades of the twentieth century. This continuity is not always obvious, though, and we might often be tempted to regard some of the early thinking about the phonograph as naïve or over-optimistic; this is the case, for example, with the suggestion that governments should encourage their citizens to record their wills instead of writing them down, which will be discussed at the end of this chapter. Either way, these questions, debates, and fantasies around the phonograph remind us that the early history of recording technologies cannot be told from a position of technological determinism: instead, how these technologies were received depended to a great extent on local and national practices and specificities, and the debates engendered in turn influenced some of the future characteristics and uses of these technologies. In Spain, indeed, the phonograph excited the imagination of Spaniards in ways which expand our understanding of how recording technologies shaped and were in turn shaped by contemporary discourses on science, technology, and modernity. Newspaper articles, poems, and plays were written by individuals who—if we go by their limited grasp of technical and design details—had probably never seen a phonograph or spent any significant time studying the technology behind it. Nevertheless, the relative abundance of such writings speaks to how powerfully Spaniards were struck by the possibility that now it was theoretically possible to record, store (presumably for an indefinite period of time), and play back sound.

Phonographs in Spain: Sound as spectacle

After the first article on the phonograph published in *La gaceta industrial*, other regional and specialized publications followed suit, typically printing translations or summaries of articles originally published in the United States or the United Kingdom without adding much by way of commentary.[5] The first opportunity Spaniards had to see a phonograph in the flesh was in September 1878 at the Ateneu Barcelonès—which, like the multiple *ateneos* operating in Spain at the time, aimed to promote the dissemination of scientific ideas among the general public, particularly the bourgeoisie. Demonstrating the phonograph was Tomás Dalmau, who was certainly not a stranger to the Barcelona science and technology circles: in 1874, he had pioneered the introduction of electricity in Barcelona, having acquired a dynamo for the Laboratory of Physics of the Engineering School of Barcelona.[6]

Dalmau's day job was in the manufacturing of scientific equipment. This is significant, because several other future pioneers of the phonograph in Spain came from similar backgrounds: as will be discussed in Chapters 3–5, some of the early recording studios in Madrid (*gabinetes fonográficos*) were set up by manufacturers of scientific equipment too, and not by musicians or show business impresarios. Manufacturers of scientific equipment like Dalmau would have had expertise in building, importing, adapting, and developing a range of equipment that could be used by schools, universities, or wealthy amateurs.[7] We can well imagine Dalmau getting acquainted with the phonograph as part of his work activities, experimenting and tinkering with it as he prepared the device for public exhibition. Remarkably, even though other individuals demonstrating phonographs in the following years and in the 1890s claimed to be Edison's "representatives" or associates, Dalmau never presented himself as such, and neither did the press reports covering his demonstrations. Moreover, Dalmau's name is nowhere to be found in The Edison Papers at Rutgers, which suggests that he acquired a phonograph privately and mainly out of scientific and not commercial interest, and that he had no formal connections to Edison.

In line with the Ateneu's aims, Dalmau's phonographic demonstration was presented as a predominantly scientific event. The event was advertised as members-only, and Ateneu affiliate Joaquín María Bartrina preluded the demonstration with a scientific talk.[8] In other respects, though, the format of the session prefigured some of the 1890s phonographic sessions, which aimed at entertaining and instructing in equal measure: Dalmau invited a few prominent local personalities and artists (including general captain of Catalonia Eduardo Blanco, tenor Rincón, and singers Rovira and Emerenciana Wehrle) to speak or sing in front of the phonograph and the audience, then played back the resulting recordings so that those in attendance could reassure themselves that the phonograph could indeed reproduce sound to an acceptable level of fidelity. As I will explain in more detail in Chapter 2, this format was adopted by many later demonstrators for two reasons: first of all, it gave the audience an opportunity to reassure themselves firsthand that the phonograph was indeed able to reproduce reality; secondly, when local singers or speakers were used, this would have provided an extra reassurance, as the audience would have been familiar with how a particular voice sounded and would therefore be less suspicious that the operator had manipulated the recording in any way. This strategy would have also provided an additional affective dimension to the experience of fidelity: the appeal of the phonograph was not in hearing a disembodied voice singing or playing a piece of autonomous aesthetic value, but rather in verifying firsthand that the phonograph worked by hearing local voices one was familiar with and could recognize easily.

The summer following Dalmau's demonstration at the Ateneu, phonographs were exhibited in two towns in the surroundings of Barcelona (Sabadell and Pont de Vilumara) during the local *fiestas*.[9] The short advertisements and notices in the local press did not mention the names of the organizers or operators, but it is likely that Dalmau was behind both events, given that no other individuals are known to have owned or exhibited phonographs in the Barcelona area at this time. Moreover, the name of Dalmau does appear in connection with a phonographic demonstration during the celebrations of La Mercé in Barcelona in September 1879,[10] suggesting that he might have organized a small summer tour of local *fiestas* in the Barcelona area to demonstrate the phonograph alongside other technologies, such as the microscope, the telephone, and electricity. If Dalmau did indeed organize a tour, this would be significant too, because it prefigures the connection that recording technologies would develop with mobility in the following years: as will be discussed later, the other two main phonograph exhibitors in Spain at this time toured the country too, and in the 1890s numerous phonographs were toured and paraded around Spain. Of course, the reason for this is that taking the phonograph from one place to another was the most practical, logical strategy for an operator to reach the highest possible number of people—but what is relevant for the early history of recording technologies in Spain is that mobility soon became a property that people readily associated with the phonograph, and both mobility and the phonograph, in turn, were associated with modernity and progress. This became more obvious in the 1890s, as will be examined in Chapter 2, but traces of these discourses can already be found in the 1880s.

In September 1880, a Mr. Espluga demonstrated a phonograph at the Teatro Jovellanos in Barcelona.[11] Espluga, a foreman, had collaborated with Dalmau in 1874 in the installation of a stall at the Plaça de Catalunya, where a telephone was publicly exhibited.[12] It is plausible that Espluga had learned about the phonograph from Dalmau, and he might have even borrowed Dalmau's device (as phonographs would be expensive and complicated to obtain at the time), or he might have been covering for Dalmau on that particular day; he is not known to have organized any other phonograph demonstrations. As for Dalmau, his last known foray into phonography took place in November 1880, at the Associació d'Excursions Catalanas—a Barcelona society organizing cultural and scientific activities.[13] It is likely that, once that Dalmau discovered and experimented with the potentialities of the first, quite rudimentary phonograph, he then felt that he had exhausted them and preferred to move on to other scientific interests—such as electricity, where he remained active throughout the 1880s. He is certainly not known to have engaged with the phonograph after 1880.[14]

Whereas the first appearance of the phonograph in Barcelona is well documented, it is less clear when the device was first exhibited in Madrid.

A newspaper in León reported in mid-January 1879 that the marquis and marchioness of Villaverde had organized a private phonographic demonstration in their home in the capital.[15] It was also around that date that the newspaper *La época* informed its readers that plans were underway to open a salon in Madrid to demonstrate the phonograph, telephone, and other modern inventions,[16] with the first advertisements appearing at the end of the month. The advertisements do not claim, however, that this was the first time a phonograph was exhibited in Madrid, which we could have expected them to do as a way of attracting the attention of audiences.[17] The sessions in the salon, which took place from eight o'clock in the evening to midnight and lasted for approximately one hour, were organized by a scientific society called Sociedad El Fonógrafo. They also exhibited a number of microphones and telephones, and were reported to use all these devices in combination in their shows, presumably to demonstrate primitive versions of an answer phone and other inventions.[18]

In February 1879, the managers of El Fonógrafo (Misters Ibarra, Berastegui, Pimentel, and Perujo) publicly announced that they had asked the Madrid municipality for permission to install a microphone and phonograph at the Teatro Real, with the aim of recording operas to be listened to from home.[19] This claim cannot be taken at face value: with tinfoil sheets being insufficient in length to allow the recording of a full opera, and with the technology still insufficiently developed to presume that listeners would derive any sort of meaningful aesthetic enjoyment from the experience, it is plausible that the managers of El Fonógrafo simply intended to make an announcement that would shock readers and procure publicity for their business. Nevertheless, the fact that they came up with the idea in the first place is significant and unusual in the context of the early history of recording technologies in Spain. At this time, Edison did still not count the broadcasting of music as one of the main uses of the phonograph. Similarly, even though Dalmau's demonstrations included music, their appeal was not so much in the music per se but rather in its ability to demonstrate that the phonograph could record, store, and play back different types of sound—not just speech. It is therefore remarkable that the managers of El Fonógrafo were able to envisage the potential of the device in procuring aesthetic enjoyment from the experience of listening to extended pieces of music; however, as is to be expected, there is no evidence that they ever managed to realize their idea. The last advertisement of the Sociedad El Fonógrafo dates from mid-February,[20] suggesting that, once the abilities of the phonograph were witnessed by Madrid audiences, their further potential was seen as limited.

The first large-scale initiative to bring the new device to audiences around the country was the tours of professional French illusionist Jean-Marie Bargeon de Viverols around Spain between April 1879 and March 1880.[21] Indeed, the scale and ambition of Bargeon de Viverols's tour were in a different league from Dalmau's or El Fonógrafo's: he was at the theaters Romea and Principal in Barcelona in April 1879,[22] Girona in May,[23] Valencia and then Madrid in June,[24] and Burgos in July.[25] There is no evidence that his tour continued in the late summer and autumn 1879; in January 1880, though, Bargeon de Viverols reappeared in the south of Spain: San Fernando[26] and Málaga[27] first, then Córdoba, Alicante, and Cartagena in February,[28] and finally Murcia in March.[29] It is rather remarkable that Bargeon de Viverols was able to obtain a phonograph and put together a show in a relatively short period of time after the phonograph was invented and the first machines were sent to Europe. The reason for this probably needs to be sought in the fact that, as a professional illusionist, he would have the expertise and resources to learn about and acquire new technologies that could be useful for his show. Compared to the limited evidence we have about El Fonógrafo and Dalmau, the tours of Bargeon de Viverols also offer greater scope to analyze how his phonograph demonstrations fitted within contemporary discourses about science and modernity/modernization, and within contemporary practices in audio-visual spectacles.

Bargeon de Viverols typically presented himself as a practitioner of thaumaturgy, Spiritism, and necromancy: he had done so before 1879–1880, and he would go back to it in later visits to Spain.[30] In his phonograph tour, however, he did not refer to the above-mentioned practices and chose to focus instead on science. A straightforward explanation is that science would have provided Bargeon de Viverols with a patina of respectability while at the same time attracting publicity for his shows at a time where science was increasingly entering the Spanish imagination. On a deeper level, though, Bargeon de Viverols's seemingly unproblematic transition between science, spectacle, and what we might term as magic throughout his career is indicative of the multifarious discourses about science that existed at this time in Spain in complex relationships with each other,[31] with the boundaries between academic and public science not always being clearly defined.[32] Bargeon de Viverols inserted himself in the context of a long tradition of scientific spectacles which, dating back to the early eighteenth century, relied on exposing audiences to apparently inexplicable phenomena, only to subsequently explain them through scientific means, with the scientific demonstration acting as a conclusion or anti-climax to the spectacle's narrative.[33] His spectacles, therefore, must not simply be seen as a way of disseminating scientific discoveries among the public, but instead be understood as a discourse on science in its own right, with its own complexities and particularities. At the same time, not everyone in the audience would have read Bargeon de Viverols's spectacles in the same way: some individuals might have attended his demonstrations with a view of getting acquainted with some of the latest scientific discoveries, while others might have been content to witness the apparently inexplicable phenomena that Bargeon de Viverols presented without worrying much as to how they came about.

Manuals written by magicians and spiritists like Bargeon de Viverols, and published or circulating in Spain during the later decades of the nineteenth century, indeed confirm that practitioners did not see their disciplines as diametrically opposed to science, and many explicitly tried to position their practices within science in different ways. For example, Carl Willmann, the author of a treatise on "salon magic" that was translated into Spanish in 1897, argued that magic had been crucial in shaping the early history of scientific research,[34] but stated that the modern magician should be aware at all times, and make it clear to audiences, that his or her tricks were nothing other than experiments in physics, optics, and mechanics, albeit conducted very rapidly so as to fool the spectator.[35] Willmann also claimed that spiritism was a superstition that *bona fide* magicians should combat.[36] Influential French spiritist Allan Kardec, whose *Espiritismo experimental* was translated into Spanish in 1904, argued that spiritist manifestations were not in contradiction with the principles of science, but thought that scientific research was not yet in a position to fully explain them following "the laws of matter."[37] He nevertheless disapproved of "fanatical spiritists" who allegedly emphasized the supernatural and downplayed the scientific aspects of their practice.[38] Scientific popularized Armando Baeza y Salvador treated animal magnetism, hypnotism, and spiritism as "mysteries" which could, however, be explained scientifically—as he tried to do in his book.[39]

The second context in which Bargeon de Viverols's demonstrations must be inserted is the optical spectacles that became central in popular culture, in Spain and elsewhere, throughout the second half of the nineteenth century.[40] The gradual introduction of these spectacles demanded that audiences develop an understanding of their visual codes, as well as an ability to follow specific reading modes to make sense of what they saw.[41] A phonograph demonstration was different from these spectacles in that the main medium was sound, and not image; however, with Bargeon de Viverols being a seasoned performer, it is entirely likely that he was familiar with optical spectacles and that he borrowed some of their conventions so that audiences could interpret his demonstrations more successfully. Bargeon de Viverols's audiences, therefore, would not have encountered the phonograph in a vacuum; instead, their experiences would have been influenced by their awareness of existing discourses around science and by their familiarity with the reading modes in operation in optic spectacles, as will be discussed subsequently.

The processes by which audiences got acquainted with the scientific discourses presented by Bargeon de Viverols in his spectacles and with the reading modes to be applied to the phonograph started before they even set foot on the venue, with the performer trying to establish his credibility as a scientist or scientific popularizer in a number of ways. Echoing Baeza y Salvador's claims in his book about magic, Bargeon de Viverols claimed in his advertisements that his show intended to expose "the lies of Spiritism" and "the false miracles of our days."[42] In his advertisements he also included an abbreviated, didactic narrative of the history of the phonograph thus far; significantly, he did not give credit to Edison for first inventing the phonograph, but rather to fellow citizen Charles Cros. In his advertisements, Bargeon de Viverols also presented himself as a member of the "Academia de Física de París" and a representative of Edison's. Both claims would have no doubt helped him establish some credibility with Spanish audiences of the time; they must not, however, be accepted uncritically. An Académie de Physique did not seem to have existed in Paris at the time, although the Académie de Sciences in Paris did have a Physics and Chemistry section; there is no evidence, however, that Bargeon de Viverols was a member. As for the magician's claim to be an Edison representative, numerous exhibitors in different parts of the world claimed to be so, but this did not always mean that they were directly employed by Edison or had been selected by him or one of his companies to exhibit the phonograph. While such exhibitors existed in the United States, each of them allocated a specific territory by Edison's company, there is no evidence that Edison set up a comparable network in Europe.[43] Among the Edison Papers there is indeed a telegram from Bargeon de Viverols;[44] dating from October 1888, shortly after Edison launched his Perfected Phonograph. In the telegram, Bargeon de Viverols explained that he had been offering phonographic demonstrations in Europe in the past few years and would now be interested in obtaining details about the recently launched new model. The telegram does not imply in any way that Bargeon de Viverols was employed by or had a substantive professional or contractual relationship with Edison's company—although it is certainly plausible that the company might have been aware of Bargeon de Viverols's activities in Europe, as one of several showmen using phonographs in their spectacles.

The content of the session themselves is known to us mostly through advertisements and not through detailed press accounts, which makes it difficult to ascertain whether the events really lived up to Bargeon de Viverols's promises, as well as how audiences responded to them. In the advertisements, the audiences were promised a chance to hear different types of sounds, mostly involving the human voice:[45]

> Reproduction of the human word, conversations with the phonograph, imitation of animal cries, military commands, arithmetic of the phonograph, cornet solos by the phonograph, comic song by the phonograph, etc.

These types of sounds (with the exception of "arithmetic of the phonograph"[46]) match what we know about other demonstrations from these years and also through the 1890s. Music, although present, was still not the predominant focus of the demonstrations, as recording technologies indeed only came to be associated with music from the late 1890s onward. Instead, Bargeon de Viverols's aim was likely to offer to his audiences a range of sounds which demonstrated the capabilities of the phonograph. From the advertisements, it is not clear whether Bargeon de Viverols recorded sound in front of the audience and then played it back or whether he availed himself of a stack of pre-recorded tin foils. The former was what Dalmau had done, and also what the demonstrators active throughout the 1890s overwhelmingly did—as a way of proving to the audience that the phonograph could indeed reproduce reality and reassuring them that the operator had not manipulated the recording before playing it back, as will be discussed in Chapter 2. However, it is likely that Bargeon de Viverols used pre-recorded plaques for at least some of his numbers: it would not have always been practical, for example, to secure the presence of different types of animals in the venue to be recorded in front of the audience. On the other hand, the program did include a section called "experience for the audience" (*experiencia para el público*), which we might presume could have involved recording and playing back the voices of volunteers from the audience and hence proving that there was nothing mysterious in the phonograph's recording processes.

Other segments in the demonstrations, however, suggest that Bargeon de Viverols (and potentially his audiences too) might have gone beyond the idea of the phonograph as merely reproducing sounds from real life. This might have been the case with the section "repetition, substitution or superposition of the sounds of the singing phonograph" (*repetición, sustitución o sobreposición de los sonidos del fonógrafo cantante*). Advertisements and accounts do not provide details as to what this section entailed, but, if we take it literally, it might have included manipulations of the device or the plaque that would allow certain sound effects to be achieved. This would have been highly unusual indeed in the context of the early history of recording technologies. Other Spanish written accounts from the period 1878–1882 and from the 1890s do not provide any evidence that demonstrators ever indulged significantly in the manipulation of sound to obtain auditory effects that were manifestly artificial. Indeed, since the main appeal of these sessions for audiences was likely to be the reassurance that the phonograph could reproduce reality, it is difficult to imagine most audiences and operators would be interested at this stage in the phonograph's potential to generate previously non-existing sounds. With Bargeon de Viverols being a pioneer in the exhibition of phonographs for recreational purposes in Spain, it is plausible that he had to operate on a trial-and-error basis, trying out different numbers that he would then reutilize or abandon depending on how successful they were with audiences. Showing or demonstrating how recordings could be manipulated might have also provided Bargeon de Viverols with a further platform to demonstrate how the apparently mysterious noises emanating from the phonograph had a scientific explanation.

But it is perhaps the segment Bargeon de Viverols called "the fireman" (*el bombero*) that best exemplifies his intentions to prove that magic and Spiritism could be explained by scientific reason. The section involved Bargeon de Viverols calling upon the death, after which "Spiritist manifestations" would be heard. Even though advertisements and accounts do not provide details, we might presume that these would later be revealed to be recordings played back by the phonograph. We might even hypothesize that this climatic moment might have been visual rather than purely auditory, with a phonograph being displayed in front of the audience, therefore providing a connection with the previously mentioned spectacles involving visual technologies.[47]

Not many phonograph demonstrations took place in the two years following Bargeon de Viverols's tours. Bernardo Rodríguez Largo, a professor of Physics at the Instituto de San Isidro (a secondary education institution), offered a talk at the Círculo de la Unión Católica in Madrid in January 1882, although notices did not clarify whether the talk included a phonograph demonstration or even whether a phonograph was on display.[48] If either of these was the case, this might indicate that some education institutions might have acquired phonographs around this time. Similarly, in February 1882 the brief for the well-known Catalan-language poetry competition Jochs [*sic*] florals stated that the poet who was judged to be "the best at singing modern life" would receive a prize consisting of a phonograph. There was no indication of the origin of said phonograph, but given that the prize was based in Barcelona, it is plausible that the organizers of the Jochs florals procured the device through Dalmau.[49]

February 1882 also saw the arrival in Madrid of Dr. Llops, the third man who, after Dalmau and Bargeon de Viverols, dedicated significant time to the exhibition and demonstration of early phonographs in Spain. Llops was presented by some Spanish newspapers as an "American physicist,"[50] but not much else is known about him. As is the case with other showmen from the period, it is not implausible that he might have exaggerated his scientific credentials or lied about his nationality. Llops's only other recorded visited to Spain took place in the period 1884–1885; he gave a tour in which he exhibited a giant microscope (*microscopio colosal*), suggesting that he specialized in shows with a scientific theme.[51]

In his first recorded phonographic demonstration at the Teatro de la Comedia in Madrid, Llops started the show by offering a short explanation about the technology itself, further establishing the scientific value of the event. He then proceeded to play back a series of numbers on the phonograph. These, again, prefigure the sorts of programs that would be offered in the 1890s, consisting mostly of vocal music or spoken word: Llops's show included a greeting to the audience, a *redondilla* (popular poetry), poetry, an excerpt from the opera Campanone, and finally a flugelhorn solo.[52] As with Bargeon de Viverols, newspapers did not mention whether sound was recorded in front of the audience and then played back in Llops's demonstrations.

Llops offered a second round of phonographic performances later the same year, this time at the theatre Príncipe Alfonso in Madrid; existing sources do not offer any other information on whether he stayed in Madrid in the intervening months. This time—perhaps after careful evaluation of what had and had not worked in the first round of performances—Llops's sessions became more akin to the phonographic demonstrations we find in the 1890s. He hired professional singers and actors who were well known locally in Madrid to sing or recite in front of the phonograph and then being immediately played back. Among those hired by Llops, bass Daniel Banquells went on to record for *gabinetes fonográficos* around 1900, which provides a further connection to the future recording industry in Spain.[53] *La correspondencia de España* and other newspapers stated that Llops's phonograph had improved capabilities by comparison to the one he had demonstrated at the Teatro de la Comedia.[54] Newspapers did not provide any further detail and we do not have any other indication of the type of phonograph Llops used other than highlighting its "quality" and "robustness." Indeed, some of these press accounts are phrased in very similar ways, suggesting that journalists simply copied Llops's publicity without offering much by way of original commentary.[55] The fact that Llops's innovations were mentioned and singled out as a significant part of his show, however, is interesting in itself, because it anticipates another development that would become significant in the 1890s, the era of the *gabinetes fonográficos*: indeed, during this time, *gabinete* owners and operators were active in introducing innovations and additions to the phonograph, and these featured heavily in publicity and advertisements as the *gabinetes* tried to establish their scientific credentials. We can presume that Llops was similarly trying to establish his credibility with potential Madrid audiences.

As with Dalmau's and Bargeon de Viverols's demonstrations, we do not fully know how audiences responded to the phonograph once they were lured into the show. The most comprehensive source we have in this respect is the account of one of the sessions published in the newspaper *El día*. The anonymous journalist approved of Llops's efforts to turn the session into a scientific demonstration and wrote approvingly that "for the first time, science took hold of that stage." He also lamented, though, that the audience found Llops's scientific explanation "too long and annoying."[56] These audience disagreements further illustrate how early demonstrations of the phonograph had to constantly negotiate their place within discourses concerning the dissemination of science: the fact that Llops and others like Bargeon de Viverols presented themselves as scientific disseminators and went to great lengths to establish their scientific credentials does not mean that this was the primary interest of the audiences that attended their shows.

After finishing his sessions at the theater Príncipe Alfonso, Llops traveled in September 1882 to the town of La Granja de San Ildefonso, north from Madrid, where King Alfonso XII and his wife usually spent the summer. Llops demonstrated the phonograph in front of the monarchs (and also at the nearby Teatro de Segovia) in what seems to be mostly a publicity campaign to raise Llops's profile; twenty years later, some of the *gabinetes* and the Compagnie Française du Gramophone upon its arrival in Spain would be similarly keen to establish connections with royalty.[57]

After Llops left the country, records of phonograph demonstrations become scarce again. An *espectáculo de fonógrafos* was open in the Paseo de Recoletos in Madrid between May and August 1895.[58] Given that it seems the sessions had a predominantly recreational scope, it might be that they were organized again by El Fonógrafo or by some of its former members individually. Even though advertisements appeared regularly in the press, no articles or commentaries were published on the installation. This is in itself interesting: it suggests that, even though discussions of the phonograph continued in Spain—as will be explained in the next section—real phonographs themselves did not excite much attention anymore. In March 1887, a "literary-scientific soirée" (*velada literaria-científica*) took place in the Madrid seminary presided by Cardinal Payá and including "experiments with the phonograph." We do not know who organized the session or operated the phonograph, but it is likely that the phonograph was acquired by the seminary for educational purposes.[59] Nevertheless, as I will discuss in the next section, speculation about the potential uses of the phonograph, which had started alongside the initial demonstrations and continued during the following years, illustrates how recording technologies captured the imagination of Spaniards in ways that both confirmed and took further their thinking about science, technology, and progress.

What are phonographs for?

Even though Spaniards did not see many phonographs in the flesh from 1882 onward, this did not stop some of them from imagining and writing about what phonographs—once properly improved and developed—should be used for. Such intended uses of the phonograph, which I discuss in this section, reveal the cultural values and properties that Spaniards attached to recording technologies at this stage. Equally revealing in this respect, though, are notions of what the phonograph was *not* thought to be for. Indeed, certain uses of the phonograph that then became commonplace in future decades, or that were indeed commonplace in other countries, were absent from discussions of the phonograph in Spain at this time. What this suggests, again, is that the roles and uses the phonograph came to acquire (namely, the recording and playing back of music) were not simply a product of technological advancement, but instead came about through cultural negotiation, friction, and change in specific contexts.

One such property of the phonograph, and perhaps the one we most readily associate with the popularization of recording technologies—is the idea that recording technologies commodified music. We commonly assume that music becomes "thinglike"[60] when it is recorded on a mass scale; from here, the concept of recorded music—connected to yet separate from that of live music—develops.[61] The emergence of the idea of recorded music as a commodity, which we might regard from our perspective as natural and straightforward, was in fact slow and convoluted. In Spain, as will be explored from Chapter 3 onward, the first signs that recorded music was being thought of as a commodity do not appear decidedly until the very last years of the nineteenth century. In the era 1877–1888, however, the idea of sound-as-commodity was still alien to debates around the phonograph; this was true of both Spain and the rest of the world.[62] Some of the reasons for this have to do with the technology itself: with few Spaniards having the financial means and the technological know-how to keep a phonograph in their own homes, there was no incentive for companies to develop the recording as a product that phonograph owners would come back to buy repeatedly. Moreover, the precarious sound quality of phonographs of the time makes it difficult to believe that Spaniards could have envisaged the phonograph as a device providing aural experiences that were pleasurable enough to become commodified. Instead, the appeal of phonographic demonstrations at this stage was likely in reassuring oneself that the phonograph was indeed capable of playing back sound to an acceptable standard of fidelity.

This means that it is also likely that no particular significance was attached to the resulting tin foils: none of those produced in Spain has survived to our day, which likely has to do with both the fragility of the material itself and the lack of commodity status. In his dictionary of Spanish historical singers, Joaquín Martín de Sagarmínaga claims that the well-known singer Julián Gayarre recorded on tinfoil for the Teatro Real, and the Teatro Real then gifted the recordings to audience members.[63] This claim has been convincingly debunked by Gómez-Montejano,[64] who argues that there is no other material or written evidence of these plaques were ever made and that, more generally, it is implausible that a well-paid, in-demand singer such as Gayarre would have agreed to submit himself to the time-consuming, precarious process of recording on tinfoil. In fact, Gómez-Montejano claims, to produce even just 100 recordings for those audience members acquiring the most expensive seats would have involved a significant amount of time and effort (let alone making recordings for all members of the audience, as claimed by Martín de Sagarmínaga). Apart from the reasons suggested by Gómez-Montejano, though, there is another, more fundamental one why it is unlikely that Gayarre ever recorded on tinfoil under the previously described conditions: at the time, the idea of music as a commodity was not developed as such, and so a plaque of Gayarre's voice (that patrons could not have played back by themselves anyway, as phonographs were normally not available for domestic use at this time) would have not hold the significance for contemporary audiences it would come to acquire later.

Also absent at this time in Spain is the idea of the disembodied voice, which was commonplace in English- and French-speaking areas at this time, and prominently featured especially in literary works. In English-speaking areas, the idea of disembodiment, according to John M. Picker, has Victorian origins and then became a ubiquitous trope among modernists (Eliot, Joyce, Mann, Faulkner, Woolf).[65] In French-speaking literature, the trope of disembodiment acquired symbolist, misogynistic undertones with Villiers de l'Isle-Adam's 1886 novel L'Eve future. The fact that the novel was not translated into Spanish until 1920 (by the avant-garde poet and novelist Mauricio Bacarisse) further supports the notion that the early reception of the phonograph in Spain was almost completely unconcerned with the idea of disembodiment. The reasons for this might have to be sought in the relatively small role that symbolism and high modernism played in Spanish literature at the time, where naturalism was still the dominant current.

On the other hand, the notion that the phonograph should be used mainly for recording and playing back music was also mostly absent in Spain at this time—as is the case elsewhere. A reason for this is, again, the poor quality of recordings, which made it difficult to imagine the phonograph as a substitute for live music. Nevertheless, some individuals envisaged, in a pioneer fashion, a future in which the phonograph was used for the enjoyment of music, as is the case with El Fonógrafo's proposal to broadcast operas from the Teatro Real, which has been discussed earlier in this chapter. A further example comes from a satirical piece about the Madrid Sociedad de Cuartetos published in the newspaper La correspondencia de España. The author suggested that listeners might as well record a few concerts of the Sociedad to listen to repeatedly at home as opposed to attending them all, as the Sociedad allegedly played the same repertoire all the time.[66] The author's suggestion must not be taken at face value, as few audience members would have a phonograph at the time, and recording full-length quartets would not be possible in the first place given the limited durations that tinfoil plaques could hold; however, the piece is significant in that it was one of the first in Spain to anticipate that the phonograph could be productively applied to the listening of music for pleasure. On the other hand, it is true that most of the phonograph demonstrations that we know of from this era included some kind of music (mostly vocal, some instrumental), but this must be interpreted as an attempt at capturing a variety of possible sounds (spoken, music, noise) that could then be played back to demonstrate the phonograph's fidelity capabilities; it is unlikely that their aim was, primarily, aesthetic enjoyment, and reviews of phonographic demonstrations certainly did not include any comments about the quality of the music or of the performance.

Writings concerning what the phonograph should be for further demonstrate the fluidity of discourses on science dissemination at this moment in time. Some scientific journals aimed at audiences who were reasonably well versed in practical science, such as the previously mentioned *La gaceta industrial*, did publish translations and other adaptations of Edison's articles and publicity.[67] However, original texts of a primarily scientific nature originally written by Spaniards are much scarcer, and indeed the only significant example is José Casas y Barbosa's chapter "Descripción del teléfono, el micrófono y el fonógrafo" in Manuel Aranda y Sanjuán's *Las maravillas de la naturaleza, de la ciencia y del arte*, intended for scientific divulgation among lay people rather than among specialists.[68] Casas y Barbosa—who had worked as a telegraphist for Correos (the Spanish national post) and had also pioneered electricity in Spain—stated that, whereas it was not difficult to envisage a multiplicity of practical applications for the telephone, this was not the case with the phonograph because "it has limited practical virtues. Nowadays, the phonograph is an admirable and very beautiful instrument for experimental and recreational physics, but we should wait to hear what his inventor says about its future."[69] Casas y Barbosa then proceeded to list several possible uses of the phonograph, mostly following Edison's ideas.

Nevertheless, most writing on the phonograph fell, strictly speaking, outside the realm of scientific writing and research. We might be tempted to dismiss such writings as mere fantasies or inane commentary—but what is relevant here is that such fantasies were often rooted, in one way or another, into broader scientific debates and concerns that had been operative in Spain for decades before the phonograph. One such concern is the climate of representation that emerged toward the end of the nineteenth century and then dominated the twentieth century.[70] In the realm of sound, this would then give rise to the concept of fidelity (that is, the notion that recorded sound was a perfect representation of live sound), which from the 1890s onward became key in the publicity and conceptualization of commercial recording technologies.[71] Notions of representation, however, were not completely new in Spain or elsewhere: indeed, they had been key in the development of image technologies, and mostly photography, throughout the nineteenth century, with eyesight being frequently posited as the best sense through which to know reality.[72] Representation is to be understood as a process rather than a one-off shift: indeed, as Riego Amezaga points out, it is not the case that Spaniards, upon seeing a photograph for the first time, would immediately accept it as a representation of reality. Instead, the mechanisms by which they came to accept photographs as representations of reality were culturally and socially constructed, with notions of social legibility, narrative rules, and conventions developing over the years from the 1840s onward.[73]

By the time the phonograph was invented, Spaniards had already developed the mechanisms necessary to recognize photographs as representations of reality. Recording technologies, however, were still to develop such mechanisms, narrative rules, and conventions. We can presume that the demonstrations of Dalmau, Bargeon de Viverols, and Llops certainly constituted an early step in such developments; however, these did not really accelerate until the 1890s, when phonographs ceased being a rarity Spaniards could expect to encounter only very occasionally and started becoming a part of everyday life—through frequent phonograph demonstrations first, and domestic phonography subsequently. However, the idea of representation—although not considered discussion of the mechanisms through which it was achieved—was ubiquitous in most of the commentary that appeared in connection with the early phonograph demonstrations and in the few years that followed. Some writings—especially those on phonographic demonstrations—extensively glossed the phonograph's ability to reproduce reality as it was, often causing great admiration among the audience, as is the case with reviews of Bargeon de Viverols's and Dr. Llops's events.[74] These and other similar sources, however, must be read critically: we should not assume that every member of the audience accepted what they heard as a true representation of reality, especially given the quality of recordings at the time and the fact that social and cultural mechanisms of auditory representation were still underdeveloped. Indeed, it is likely that journalists were simply repeating what Bargeon de Viverols and Dr. Llops claimed in their publicity without examining it critically.[75]

Most interesting, though, is the fact that, in the minds of *fin-de-siècle* Spaniards, it was not just *audible* reality that the phonograph was thought to reproduce. Commentary also abounded on the phonograph's alleged ability to reproduce what was real but not audible. There are two distinct categories here, the first of which concerned the phonograph's ability to preserve and play back the voices of individuals who had since died. This was not a trait unique to Spain: indeed, it was widespread in commentary throughout the world in the early days of the phonograph, with Edison's own publicity suggesting that the phonograph could be used to record the voices of one's close family and friends, so that they could be played back after they were dead. In Spain, writers singled out prominent public speakers and politicians whose voices were deemed worthy of preservation so that future generations could learn the principles of rhetoric and public speaking,[76] and poems were written glossing the potential of the phonograph to preserve the voice of the dead.[77]

The second category of the inaudible concerns thoughts and beliefs which would be expressed in private but never in public—or which would never be said out loud in the first place for fear of consequences to one's social status. This second category, according to Douglas Kahn, is derived from the first: it was indeed the idea of the phonograph giving access to the voices of the dead that allowed the notion to develop that the phonograph could record "the previously unrecordable, the technologically inaccessible regions of consciousness or the mysterious."[78] The notion of the phonograph being able to reveal the unknown or hidden can also be tied back to Leibnizian understandings of the machine that spread across Europe from the eighteenth century onward: the machine was thought to not just be able to measure and quantify the university, but also to express, by virtue of complex mechanisms, what was otherwise unknown or inaccessible.[79]

It is this second category of the inaudible which took on particularly distinctive features in Spain, with notions of the inaudible being intimately connected to country-specific concerns and debates regarding the separation between the public and the private/domestic. Much of the Spanish commentary on the phonograph at this stage suggested that its ability to record the inaudible truth (regardless whether it was spoken out at any point or not) could potentially be disruptive to sexual and family mores. Writers suggested that the phonograph could ruin marriages by revealing whether each of the partners really loved the other, and in 1879 a poem was awarded a prize at a local Catalan-language competition in Barcelona about a man who proposed to a woman and, after talking to her parents, he left behind a phonograph in their home to record what they really thought about him.[80] In the convoluted political climate of the Bourbon Restoration—with the Liberal and the Conservative party alternating in the government in a controlled process and local leaders ostensibly manipulating election results, which left many disenchanted with constitutional monarchy—it was not surprising either that politics became another frequent topic of writings focused on the phonograph's ability to tell the truth. It was repeatedly suggested that the phonograph could be used to record politicians saying what they would have never dared say in public to their electors.[81] One commentator even wrote that the phonograph had the ability "to play back the opposite of what has been recorded."[82] This can be read as a commentary on the poor quality of phonographic reproduction, but also as another version of the idea that the phonograph was always able to get to the truth and do away with hypocrisy and lies: indeed, the potentially disruptive quality of the phonograph in this regard was almost invariably portrayed in a positive way, thus turning the phonograph into an ally in Spain's advance toward modernity—not just scientific and technological, but also political and social.

Apart from newspapers and magazines, much of the commentary about the phonograph as a speaker of otherwise unspeakable truths can be find in theatrical and musical plays. This is consonant with the status of theater at the time as the preferred pastime of many Spaniards from all social classes, with authors often seeking to attract audiences and compete in a crowded market by integrating current topics and debates in their plots. Critique or full-fledged satire was rare, though, and allusions to recording technologies in these early plays must be regarded as superficial commentary on current issues rather than as thorough critical engagement. It is likely that such commentary, however, resonated with audiences, mirroring some of their thoughts about recording technologies and in some cases exciting their imagination. Plays featuring phonographs or gramophones of different sorts continued being written until the 1910s; these included Ruperto Chapí's *zarzuela El fonógrafo ambulante* (1899), discussed in Chapter 2.

Among the plays written in the 1880s, *El fonógrafo: invento en un acto y verso* (1885), with a libretto by José Castillo y Soriano and music by Isidoro García Rossetti, engaged with recording technologies to the greatest extent. It also the one which received the most publicity in its time, in part presumably because it was the first focus on recording technologies, but also because it marked García Rossetti's comeback to the stage after an absence of some years.[83] *El fonógrafo* received one single review, which was positive,[84] and there is no record that it was performed again after it first run, but we should not assume from this that it was a failure: this was indeed rather common even for plays with a certain commercial success in a climate of rapid change and competition. García Rossetti's music has not made it to our days, so my analysis will focus solely on the libretto.

Stage directions throughout the play suggest that Castillo y Soriano did not intend for a real phonograph to be used, confirming that he was primarily interested in presenting widespread tropes about recording technologies for comic effect, but not so much in exploring how said technologies could offer new dramatic possibilities. Instead, it is likely that some sort of prop manufactured specially for the occasion was used, and that Castillo y Soriano himself and his audiences were not familiar with how a phonograph would have looked. In one of the scenes—which will be further discussed subsequently—Próspero, the protagonist, announces that he is about to go for a "paseo fonográfico" (phonographic stroll) with a phonograph fitted inside his pocket. The libretto never explains how this is meant to be achieved, or whether Próspero's attempts are meant to have comic effect, which supports the notion that a smaller prop might have been used.

Another indication that Castillo y Soriano did not intend for a real phonograph to be used is that he wrote in the libretto that there should be an indefinite number of people ("voices and choirs of the phonograph"—in the original, *coros y voces del fonógrafo*) off-stage, singing or speaking whatever words or music were meant to come out of the phonograph at any given time. This would have likely enhanced opportunities for comic effect, as a real phonograph might have been too soft and rudimentary at this time to convincingly play back a range of sounds to a large audience. The last scene, in which the phonograph is supposed to play back different kinds of sounds (birdsong, bells, thunderstorm, hymns, etc.) would likely provide an opportunity to the off-stage troupe to demonstrate their vocal and comedic abilities.

Instead, *El fonógrafo* engaged with widespread tropes in the Spanish imagination about recording technologies by inserting them into what would be typical plot elements (adultery, mistaken identities, mild critique of bourgeois morals) in commercial theater. The play opens with Próspero, a middle-aged bourgeois, returning to Madrid after traveling to Paris and acquiring a phonograph there. In Paris, Próspero also had an affair with a Madame Fiffí; at the same time, while he was away, his wife Rosa was being courted by his friend, Pepito. It is these adulterous affairs that the phonograph disrupts, to comic effect. Rosa learns about Madame Fiffí after hearing a recording that Próspero made of her voice in Paris, and Pepito, determined to declare his love to Rosa, has to resort to communicating through signs after he spots the phonograph, for fear of being recorded and found out by his friend. These set-ups are consonant with typical plots in commercial theater, but also with the nascent view of the phonograph as a machine which through technological means, was able to speak and reveal what no one dared say in public. The notion of the phonograph as representing those aspects of reality that some would like to keep hidden is also obvious in the "paseo fonográfico" scene, with Próspero saying: "Oh, how I will laugh when I hear back home everything that runs here and there. The phonograph is, without a doubt, the newspaper of the future. It is the impartial echo of a country's opinion."

Two more plays were premiered in 1886 and 1887, after the initial interest in phonograph demonstrations had subsided. *¡El arte del toreo!*, a *revista* with libretto by Ricardo Monasterio and Julián García-Parra and music by Manuel Nieto (1886), envisages a future in which the voices of famous contemporary bullfighters are preserved in the mid-twentieth century thanks to the phonograph. *El bazar H*, with libretto by Calixto Navarro and Manuel Arenas and music by Manuel Fernández Caballero (1887), revisits the idea of the phonograph as a machine that can reveal hidden or difficult-to-access truths. The play is a loosely tied collection of sketches all set in a department store in which several European countries (France, Germany, Spain, Russia, etc.) are anthropomorphized and sell products according to their alleged strengths (e.g., Germany in war, France in fashion and jewelry, etc.). The phonograph features in one sketch: it repeats not what has been said in front of it, but rather what the person recording the message really has in mind—for example, a politician is found out not to really want the best for his country but only think about personal advancement, and a wife is found out not to love her husband.

Although falling outside the chronological scope of this chapter, there is another text that extends the idea of the phonograph as witness of the truth and further points at how scientific discourse extended beyond purely scientific venues and publications: I am referring to Publio Heredia y Larrea's *El testamento fonográfico*,[85] dating from 1895. Heredia y Larrea had had by that point a thirty-year career as a clerk in the Ministry for Justice and as a judge. The book is significant in that it follows through in detail one of the ideas opened up by the phonograph's potential to tell the truth: as indicated by the title, Heredia y Larrea explores the implication of implementing the "phonographic will," that is, a will that is recorded instead of written down. Heredia y Larrea was careful to point out that he did not think scientific progress should be adopted blindly,[86] but rather introduced cautiously.[87] He claimed that the phonograph would soon not only "broaden and facilitate" social relations between people but also "transform them completely"[88]—although, from today's perspective, it is clear that sound technologies have transformed social relations in a very different way from what Heredia y Larrea anticipated.

The phonographic will would, according to Heredia y Larrea, eliminate ambiguities and would also help prove conclusively that the will had not been falsified, as one could easily determine whether the recorded voice belonged to the person in question or not.[89] Intriguingly, he also cited a further virtue of recording as opposed to writing down a will, which recalls the earlier fascination of Spaniards with the phonograph being the voice of the dead: he argued that listening to a will being read out on a phonograph, as opposed to reading it on paper, would cause a psychological effect of surprise in participants which would add an element of solemnity and seriousness to the proceedings.[90]

Conclusion

The initial interactions of Spaniards with Edison's phonograph—be it under the form of responses to actual demonstrations or fantasies about how recording technologies might evolve in the future—helped shaped at least two notions which became central in subsequent decades. Firstly, it was in this era that phonograph demonstrations started acquiring some of the characteristics and formats that would become standardized during the 1890s, as Edison's Perfected Phonograph reached larger groups of Spaniards. From the 1878–1882 sessions, the notion emerged that the main appeal of listening to the phonograph was to reassure oneself—and being reassured by others too who shared the experience—that recording technologies were indeed able to reproduce reality to an acceptable level of fidelity, thus making the phonographic demonstrations and its significant entertainment component part of the multiple discourses on the dissemination of science active in Spain at the time. In the 1890s, as will be discussed in the next chapter, while many of the basic characteristics and aims of the original phonographic session remained the same, they were developed further, incorporating a significant sociability component; from the late 1890s, commercial phonography replaced these with new ways of listening and consuming music in a private, domestic space.

Secondly, the notion also developed that the phonograph was able to reproduce auditory reality as it was—and, in some cases, as it should be. Whereas at the time this remained in the realm of the speculative, as most Spaniards would not have regularly interacted with recording technologies, the notion was surely at the basis of the various cultural processes and negotiations which, from the early 1890s and more decidedly from the end of the decade, resulted in commercial recordings being generally accepted as a true record of reality.

2

Traveling phonographs, 1888–1900

The launch of Edison's Perfected Phonograph in 1888 opened the door for many Spaniards of all social classes to familiarize themselves with recording technologies throughout the ten to fifteen years that followed. During this time, phonographs remained a rarity in Spanish homes (and they would stay so until the late 1890s): instead, Spaniards got to see and hear talking machines predominantly though itinerant public demonstrations and in phonographic salons. In these, phonographs were exhibited and demonstrated in front of an audience, but they were typically not sold.

The places and spaces where phonographs were heard at this time shaped and redefined the roles that recording technologies occupied in discourses around science, technology, modernization, and national identity. Itinerant demonstrations reconfigured the phonograph as a mobile object, and this allowed the device to be inserted into *regeneracionista* discourses that connected travel and mobility with progress and increased interconnection of the various provinces and regions in Spain. Phonographs did not only move geographically, but also metaphorically across social classes, with demonstrations taking place in a range of spaces of sociability attended by different echelons—from *ateneos obreros* dedicated to the education of the working class and *salones fonográficos* which exhibited the new device at prices affordable to the urban masses, to *casinos* attracting the financial elites of the country and private gatherings at aristocratic homes. These different social groups and venues had different understandings of science, technology, and modernization, which resulted in discourses around recording technologies becoming more multifarious and diverse.

Another, more practical consequence of the introduction of the traveling phonograph is that it consolidated the "phonograph demonstration" format that I introduced in the last chapter. Although primary sources are not always unproblematic in terms of ascertaining what exactly went on during these demonstrations (as will be discussed in the course of this chapter), they certainly show commonalities in terms of what was recorded and played back in front of the audience, in which order and how the whole event was framed and presented. From this, some conclusions can be drawn as to how audiences would have perceived and read these demonstrations and, consequently, how they might have conceived of recording technologies. While the focus was still

Inventing the Recording. Eva Moreda Rodríguez, Oxford University Press. © Oxford University Press 2021.
DOI: 10.1093/oso/9780197552063.003.0003

predominantly on displaying and explaining the scientific capabilities of the phonograph and particularly its fidelity, new ways of listening to recorded music started emerging during this era too: indeed, as I will discuss later in this chapter, there is evidence that some audiences and operators were starting to appreciate recorded music as an aesthetic object, rather than a mere scientific novelty. In this chapter, I first discuss the introduction and first notices about the Perfected Phonograph in Spain in the years 1888–1890, before demonstrations properly started, and then analyze the issues that the sources pose in researching this historical period. I subsequently move on to discuss phonograph demonstrations and *salones fonográficos* in the years 1890–1905, focusing on the three issues outlined previously: the mobility of phonographs; the spaces of sociability where phonographs were exhibited; and the phonographic session as a format.

From speculation to demonstration

In Chapter 1 I discussed how real phonographs were rarely seen in Spain after 1882 but were still written about quite frequently in newspapers, the scientific press, and stage works. Similarly, once the invention of the Perfected Phonograph was announced in 1888, Spaniards spent at least two years writing about it before they could see one. The reason for this delay must probably be sought in the fact that Edison and his companies did not regard Spain as a particularly interesting market at the time (and indeed, as will be discussed in Chapters 3–5, even when commercial phonography stated to develop, the *gabinetes fonográficos* operated independently from and undisturbed by Edison). On the other hand, while individual Spaniards could have independently acquired a phonograph in the United States or the United Kingdom, this might have been too expensive and time-consuming for them to try.

Edison first announced the invention of the Perfected Phonograph in an article published in the *North American Review* in June 1888,[1] and the first commercial wax cylinders were manufactured the following year. News of the phonograph soon reached Spain, with the first documented Spanish translation of Edison's article dating from as early as August 2, 1888 in the Barcelona newspaper *La publicidad*.[2] On August 23, a note in the Madrid newspaper *La justicia* announced that the phonograph would soon be exhibited in London.[3] Other translations and adaptations of Edison's article were published subsequently in both general newspapers and the scientific press, most often with no or minimal commentary,[4] and reports about the launch of the new phonograph in the United Kingdom also eventually reached Spain.[5] Even though Barcelona's Exposición Universal—showcasing global innovations in industry and applied technology—took place between May and December 1888 and therefore overlapped with the first months of the new device's life, there is no evidence that the phonograph was exhibited there, and neither the Exposición's catalogue nor other documentation mention the phonograph at any point.[6]

The first demonstration of a Perfected Phonograph in Spain, however, might not have taken place until as late as 1892. Very few phonograph exhibitions at all are documented between 1888 and 1891. In January 1890, Pedro (Peter) Reid, an English businessman living in Puerto Orotava (Tenerife), offered a phonograph demonstration to his neighbors.[7] There is no indication in the newspaper advertisement whether Reid's phonograph was a Perfected one, but it claimed that this was the first time a phonograph had been seen in the Canary Islands. This might suggest that Reid would have only recently acquired the device, which might therefore have been one of the newer, Perfected models—however, this cannot be conclusively proven. Charles Kalb, an agent of Edison's, was offering demonstrations in Barcelona and Girona shortly after that—in February and March 1890.[8] One newspaper reviewing the events, however, claimed that he was not exhibiting the "latest Edison model," but older ones.[9]

After these isolated instances, it was only from 1892 onward that phonograph demonstrations picked up again for the first time since 1879–1882. We know that these new demonstrations used Edison's Perfected Phonograph either because reviews and advertisements mention it specifically, or because they talk about wax cylinders and not tinfoil plaques. While commercial phonograph manufacturing thrived in neighboring France throughout the 1890s, evidence suggests that most Spanish demonstrators used Edison rather than French models, as they often named Edison in their publicity or in the title of their shows or establishments (e.g., Salón Edison), or even claimed to have direct connections with him, as will be discussed later. There were a few exceptions who used French phonographs: a demonstrator called Lucio García Leal presented in 1897 in Madrid and Barcelona the phonograph invented by Frenchman Henri Lioret (commonly known as the "Lioret phonograph") two years earlier.[10] A non-commercial CD, currently held by the Biblioteca de Catalunya, containing transfers of wax cylinders owned by Spanish early recordings collector Mariano Gómez-Montejano also features a few Lioret recordings, which suggests that these, as well as Lioret's phonographs, might have also been present in the Spanish markets during these years, albeit more limited than was the case for US models.[11]

Sources for phonograph demonstrations

Although the phonographic demonstrations which took place throughout the 1890s and into the early years of the twentieth century are crucial to understanding the early history of recording technologies in Spain and elsewhere, the sources we have for their study are incomplete and fragmentary. A reconstruction of what the audiences of these demonstrations would have been able to see and, more importantly, hear, leaves us with more gaps than certainties, and, as a result, our understandings of how they might have conceptualized these events is limited too. For example, I have not been able to find photographs portraying any phonographic demonstrations from these times, which would have offered important clues as to how Spaniards might have responded to phonographs as they heard them. More importantly, no trace of the recordings themselves has survived. As with recordings from the earlier demonstrations during the 1880s, this might have had to do, in the first place, with the fragility of wax cylinders, but also with the fact that operators and audiences would have likely given little material or cultural significance to the physical recordings (as opposed to the phonographs or the experience of listening to them). Indeed, as I will argue later on, the phonograph demonstration as a genre at this stage was mostly built on the appeal of seeing and hearing the phonograph and reassuring oneself that the device was capable of capturing and playing back sound. In a context in which phonographs were still a novelty, being able to listen to the same recording again and again is likely to have been less appealing to audiences. It is therefore likely that wax cylinder recordings were produced ad hoc and discarded afterward, either immediately or after a few uses.

In the absence of visual and audio sources, we must rely on written accounts to reconstruct phonographic demonstrations—mostly advertisements and reports in the press, as well as programs of municipal *fiestas* in which phonographs were exhibited and the minutes of some associations which hosted phonographs. These accounts can only give us a partial, mediated impression of the events themselves, particularly in which concerns the sensory aspects that would have no doubt been key to the experience. Most of them are also rather brief and ephemeral by nature, intended, implicitly or explicitly, as publicity for the phonograph itself or for one of its demonstrators: we should, for example, be reasonably skeptical of any demonstration reviews claiming that the phonograph's fidelity was so remarkable that the audience were unable to distinguish recorded sound from reality[12]—although, on the other hand, we should remember that an audience who encountered the phonograph for the first time might have been impressed and surprised enough by the realization that reproducing sound was indeed possible so as to accept, at least initially, the resulting sound as a perfect reproduction of reality. Other accounts are more informative, but they rarely delve into what might have been the typical experience of an audience member: indeed, whereas literary and non-fiction writers in other cultures (particularly English- and French-speaking) recreated at some length the experience of listening to the phonograph—more often than not with sinister or uncanny undertones—the same is not true of Spanish literature. Traveling phonographs, though, continued making appearances in a few commercial theatrical plays and *zarzuelas*, the most conspicuous example being Ruperto Chapí's *zarzuela El fonógrafo ambulante* (1899), which will be briefly discussed in the course of this chapter.

The sources we do have confirm that the themes of mobility and sociability were key in how phonograph demonstrations were presented and presumably perceived by Spaniards. Phonographs were portrayed arriving and leaving towns, transported by individuals who seemed to be constantly on the move. Once in a locale, they were exhibited in public or semi-public venues, from fair stalls and ad hoc salons to cafés, theaters, and private clubs, all of them hosting distinct social groups. Researching contextual detail about the phonograph's travels, spaces, and audiences can certainly help us understand how sound technologies became embedded within discourses about mobility and sociability, while delving into the formats of phonographic sessions themselves further illuminate these two categories.

Traveling phonographs

An examination of the press repositories Hemeroteca Digital, Prensa Histórica, and Memòria Digital de Catalunya,[13] which hold digitized newspapers published in all of the provincial capitals of Spain and some of the main towns in the period under study, as well as the collection of fair (*fiestas*) programs held at the Biblioteca Nacional de España, reveals some 250 instances of public or semi-public phonograph demonstrations between 1892 and 1905. The real number of demonstrations would have no doubt been higher, probably by orders of magnitude. An "instance" here can refer to a one-off demonstration or to a series of demonstrations organized by the same person in the same place. It was common for demonstrators to spend several days in the same city or town and to offer several demonstrations within that time frame, often actively responding to demand and modifying their schedule accordingly, so that the actual number of demonstrations carried out by the same person in a place is often difficult to quantify.

Evidence from the existing press accounts suggests that demonstrations peaked in the years 1895–1900 and then started a decline which accelerated after 1903. This is relatively straightforward to reconcile with we know about the development of domestic phonography, which will be discussed in the subsequent chapters. Domestic phonography developed in Spain from the years 1896–1897 and peaked around the period 1899–1901 at the hands of the so-called *gabinetes fonográficos*. These developments meant that by around 1900 more middle- and upper-class Spaniards were able to acquire their own phonograph and cylinders, and so public demonstrations might have lost their appeal among these social groups—while demonstrations and salons still found a market among the working classes. Both the *gabinetes* and public demonstrations and salons went into a steeper decline from 1903 onward, as multinationals like Gramophone settled in Spain, introducing a broader range of products at cheaper prices.

The preceding conclusions, though, must not be regarded as definitive, as they rely on rather incomplete newspaper records. Not all demonstrations were advertised or reported about in the press, and it is likely that newspaper decisions on whether to report or not on them changed over time, with the phonograph attracting more attention from journalists in its early days and then losing novelty value. An article published in the Catholic magazine *El grano de arena* in 1905 suggests that this might have indeed been the case: the writer claimed that phonograph demonstrations were now taking place every day in the Minorcan city of Ciutadella.[14] If we look at the Minorcan press from the time, though, phonograph demonstrations were reported much more sparsely than that.

Most of the phonographs exhibited from 1892 onward and up until the mid-1900s were "traveling" phonographs. As discussed in Chapter 1, mobility already featured in some of the first phonograph demonstrations in Spain between 1878 and 1882, notably with Bargeon de Viverols but also, to a lesser extent, with Dalmau and Dr. Llops. It is likely that, in the 1890s, phonographs became mobile first and foremost because of practical reasons. A demonstrator could maximize his or her profits by covering one or a few provinces instead of a single city or town (where the market, outside Madrid and Barcelona, would be by force small). At the same time, it is likely that audiences would be more likely to attend the show if they knew that the device would be in their town for a limited time only: indeed, exhibitors often chose to emphasize in their publicity that they would be leaving the place in just a few weeks or days.[15] Throughout the 1890s, therefore, mobility became more intimately connected with recording technologies than had been the case so far, with phonograph demonstrations and the phonograph itself being often read under the light of prevailing discourses which connected mobility, national identity, and modernity, and the phonograph in turn influencing such discourses. This is rather unique in the global context, and makes Spain a prime case study to examine how recording technologies were shaped by national and local forces: not only because discourses about mobility and modernization were specific to the cultural context, but also because the focus on traveling phonographs itself was a unique Spanish trait too. In the United States, for example, "coin-in-the-slot" (i.e., stationary) phonographs were the norm, and phonograph demonstrations were more tightly regimented than was the case in Spain: Edison indeed allocated authorized demonstrators to specific regions; these demonstrators would then be briefed and encouraged to share good practice at annual conferences.[16] There is no evidence that Edison and his companies ever wanted to impose the same kind of control on the Spanish market (even though, as I will discuss later, some of the demonstrators were indeed Edison representatives): instead, demonstrators and *salón* owners operated independently from one another.

Phonograph demonstrators in 1890s Spain were indeed a diverse group. Little is known about most of them and about their motivations to pioneer recording technologies in Spain. It is likely that some were attracted by financial gain: an advertisement of the Edison branch in Paris in the Madrid newspaper *El imparcial* in July 1895 promised that an investment of 500 francs in a phonograph would result in gains of 50 to 100 francs per day; this was between ten and twenty times higher than a French manual worker could expect to earn at that time.[17] Others might have been genuinely interested in disseminating the latest scientific discoveries among their compatriots. Indeed, records suggest that some demonstrators limited themselves to occasionally organizing demonstrations in their own and neighboring towns, suggesting that they did not earn a living from the phonograph.[18]

Some of the phonograph demonstrators were—or claimed to be— representatives of Edison's and other companies in Spain. These demonstrators were indeed among the first to tour the country, with Charles Kalb followed by a Mister Natini from the French company Demeyer two years later.[19] Company representatives often focused their activities in Madrid, Barcelona, and the neighboring populations, and they only rarely ventured into the provinces. Nevertheless, there is some evidence that they were mobile too within cities, suggesting that from a rather early stage recording technologies and mobility were interlinked in significant ways. For example, a Mister Sears and Mister Warring, who introduced themselves as Edison's agents, claimed to be the first to exhibit a Perfected Phonograph in Madrid in autumn 1892, and toured several theaters while there, including Comedia and Apolo.[20] Demonstrators' claims that they were Edison's "agents," "representatives," or "disciples"[21] must not be believed uncritically. For example, the Edison papers contain no correspondence or other documents mentioning Kalb, so it is unclear that he was indeed an appointed representative or had links with the company. Some of them might have claimed that they had connections to Edison in order to boast their credibility, assuming that, with Edison's companies showing little interest in Spain, it would be difficult for them to be found out.

Many phonograph demonstrators were funfair professionals, who often had experience with a range of audio-visual spectacles such as the magic lantern, phantasmagoria, X-ray, *cuadros disolventes*, and—towards the end of the period under study—the cinematograph. These types of spectacles would often be exhibited side-by-side with the phonograph.[22] These phonograph demonstrators were often the most mobile, or at least those whose mobility patterns are more obvious to us; indeed, mobility would have probably been a key part of their professional life well before the arrival of the phonograph. A demonstrator by the name of Lorenzo Colís was active mostly in the Aragon and La Rioja regions, in Northern Spain; this would have likely given him access to a reasonably large and broad variety of audiences without steering too far away from his center of operations.[23] A Mister Bellán was active in Catalonia and the Balearic Islands at various points between 1895 and 1896,[24] a Mister Gil and Mister Casaña toured together the area of Navarra in late 1896.[25] Remarkably, at least two individuals active in demonstrating phonographs throughout Spain went on to open their own *gabinete fonográfico* at a later stage. José Navarro Ladrón de Guevara, a clockmaker, was active demonstrating phonographs in the Levante area in the last years of the nineteenth century,[26] and after 1900 he opened a *gabinete* in Madrid (operating under the names of José Navarro and La primitive fonográfica) and finally, under Fábrica de Grafófonos, became a reseller for other brands. Frenchman Armando Hugens was active in demonstrating the phonograph from at least 1894 onward, mostly in Andalusia, Madrid, and the Balearic Islands.[27] He also famously installed a "pay-per-listen" phonograph in the arcades of the Teatro Apolo in Madrid[28]—the most popular *zarzuela* theater in Madrid at the time, whose arcades typically hosted displays of new technological gadgets and other curiosities. Hugens went on to found the Sociedad Fonográfica Española Hugens y Acosta in 1896—the first and longest-lived Spanish *gabinete*. In some respects, compared to his fellow demonstrators, Hugens was in a category of his own; there is no doubt that—unlike other demonstrators and later *gabinete* owners who dabbled in recording technologies for a short time only—Hugens was truly committed to the phonograph. From an early stage, he imported a number of improvements and innovations, he was using Bettini diaphragms from at least 1896, and he emphasized the scientific nature of his demonstrations over pure entertainment, contributing more broadly to the dissemination of the phonograph in Spain with his 1893 translation of A.-Mathieu-Villon's opuscule *El fonógrafo* in 1893.[29]

Hugens, Navarro, Colís, and the rest of the previously mentioned funfair professionals presented their phonographs in a fully itinerant manner, leading public demonstrations in one place for a few days (or sometimes weeks) before moving on to the next town. Other operators stayed in specific places for longer, but they were also mobile to some extent. The first documented examples of a more-or-less permanent phonograph on display include the Fonógrafo Soley in Barcelona in summer 1893[30] and then La Equitativa in Madrid in January of the next year.[31] These were followed by a rather well-publicized installation in Madrid's Calle de la Montera, organized by an impresario called Mister Pertierra.[32] Little else is known about Pertierra; his installation remained open, on and off, for several years. During the summers, the Madrid premises closed and Pertierra toured the provinces offering phonograph demonstrations.[33] Such stationary installations became more widespread from approximately 1895 onward. Most were located in Madrid and Barcelona; in the provinces, on the other hand, *salones* were open in a city for several weeks or month before moving to another, therefore combining elements from stationary fixtures and itinerant demonstrations. An installation called "Salón Exprés" (or "Expréss" in some of the sources), managed by a man called Mariano Castañera, was active in Northern Spain during 1897 and 1898, including the Basque Country, Castile, Navarre, Aragon, La Rioja, Catalonia.[34]

Finally, other demonstrators operated outside commercial circuits: they were individuals who had bought a phonograph for their own personal use or who worked at an educational institution who had acquired one, and occasionally organized demonstrations in their own community.[35] One such individual was Ignacio Figueroa y Medieta, Marquis of Tovar, who was active organizing phonograph demonstrations among the Spanish aristocracy and royalty, including Queen María Cristina, in 1897.[36]

The travels of phonographs and their demonstrators soon became an important pillar of how such demonstrations were written about in the press. This is significant, because—unlike with other inventions such as the train—mobility was not central to the phonograph's capabilities for recording and playing back sound: writers might have chosen to leave mobility out of the picture altogether and still extoll the new device as a significant scientific innovation. The reason why they did not do so is likely that mobility strengthened the phonograph's status as a symbol of modernization, by resonating with the discourses that have been discussed in the introduction. In practice, this meant that the arrivals and departures of operators (and not simply the demonstrations themselves) were duly reported in local newspapers, and that such notices were typically pervaded with verbs denoting mobility or immobility. Phonographs were presented as arriving in, departing from, or traveling toward specific locales; sometimes, they were also reported as having been installed (*instalado*)—presumably for a short time before they became mobilized again.[37] Notices about phonographs were often printed side by side with others referring to the mobility of people, ships, trains, or merchandise. On October 28, 1898, the Galician newspaper *El Eco de Santiago* reported about a phonograph being installed in the Rúa del Villar side by side with two other notices about local doctors leaving for various destinations;[38] other newspapers published comparable examples.[39] Newspaper evidence even suggests there was an expectation that phonographs should travel to people, and not the other way around. In 1897, funfair impresario Adolfo Fo exhibited a phonograph in the center of Alicante, and he was allegedly forced by popular demand to take the device to the neighborhood of

Benalúa for a few days (instead of Benalúans traveling to the city center, which would have been fully doable given the modest distances involved).[40] A phonograph operator visiting a town to offer a series of demonstrations in a theater or hotel would similarly often be expected to engage in a short tour of the place, offering several demonstration in social clubs or at private homes.[41]

Even though these notices were rarely elaborated further and therefore do not allow us to fully comprehend how Spaniards might have conceived of the phonograph's mobile qualities, the very significant presence of mobility tropes in them makes it reasonable to consider phonographs within the context of the previously outlined discourses about mobility present in Spain at this time. The idea that the phonograph's mobile nature was crucial to give remote provinces and towns a taste of the modernization processes that the biggest metropolitan centers would have been undergoing is certainly explicit or implicit in many of these notices. Many advertisements and reports chose to highlight that this was the first time that a phonograph was exhibited in a particular city or province. Importantly, such claims sometimes contradict each other or other evidence, with several demonstrations being labeled as the first in a particular province or city.[42] They must therefore not be taken at face value, but they are significant in that they suggest that the arrival of the phonograph was widely seen as a powerful symbol of modernity physically arriving to faraway corners of the country.

The phonograph, though, was allegedly not welcome everywhere—although such accounts, appearing both in the press and in theatrical and satirical works, must also be examined critically. Several articles and short stories were published in the Spanish press during these years, describing how peasants and farmers in remote parts of Spain reacted negatively to recorded sound and became scared that the devil or another supernatural being was inside the device.[43] Importantly, few details were typically given as to when and how the episode took place, and the accounts also tend to be remarkably similar to each other: it is therefore highly possible that not all of these articles were completely truthful and were instead part of a newly developing journalistic genre. Nevertheless, they are significant in terms of what they reveal about attitudes toward modernity and recording technologies. Indeed, science-embracing *regeneracionistas* regarded "illiteracy, lack of education, and superstition" as some of the main obstacles to their efforts[44]—and remote locations outside the cities were regarded as particularly vulnerable to such issues, therefore potentially compromising the spread of modernization throughout the country. These newspaper accounts resonate with these discourses, as they portrayed rural populations as superstitious and unable to understand the benefits of modernity. Literary and theatrical works also echoed the trope of the rural, superstitious Spaniard unable to accept the phonograph for what it was. The 1898 *zarzuela El paraíso perdido*, with a libretto by Jackson Veyán and Gabriel Merino, presents a phonographic demonstration where locals listen to the phonograph "with idiot-like faces." The demonstration did not have a particularly crucial role in the plot and was simply included for comic effect, which suggests that the trope had by then become widespread among urban audiences who would immediately be able to recognize the set-up.

Ruperto Chapí's and Juan González's zarzuela El fonógrafo ambulante (1899) provides an opportunity for a more sustained study of how recording technologies might have found themselves in the center of differing or contradictory discourses on science, technology, mobility, and modernization. Chapí is well known as one of the most successful zarzuela composers of his time; González, on the other hand, is more elusive. He is not known to have written any other zarzuelas and on the occasion of the premiere, one Madrid newspaper claimed that the librettist ran away from the theater as soon as the audience started to call his name at the end so that he could take a bow.[45] A later review in Palencia intimated that the author was "a high-up, well-known aristocrat" who hid under a pseudonym.[46] If this was indeed the case, González might have been the Marquis of Alta-Villa, a diplomat, singing teacher, and phonograph enthusiast. Alta-Villa's engagement with recording technologies took on in the gabinetes era, as will be discussed in Chapter 3, but it is plausible that by the late 1890s he knew the phonograph and its audiences well enough to portray them in the libretto to the extent that he did.[47]

Like El fonógrafo, discussed in Chapter 1, El fonógrafo ambulante was eminently commercial theater, aimed at providing entertainment rather than at making sophisticated social, cultural, or political arguments. Therefore, I take it as my starting point that the ideas about recording technologies expressed in the zarzuela would have resonated with a significant part of the audience. Perhaps it would have amplified or satirized in some ways the audiences' own perceptions of recording technologies, but it is less plausible that it would have challenged them in a fundamental way, as this is not what zarzuela audiences would have expected of the genre. Such discourses about recording technologies are subsumed in a view of Spanish society which is entirely typical of zarzuela at this time: the Spanish pueblo (i.e., the working and lower middle classes) are portrayed as devout, patriotic (but also proud of their regional background), and family and community oriented. They are also resourceful in overcoming any problems that life might through their way and generally distrustful of government and state institutions—but never so much that they might seriously consider organizing against them.

Within this framework, El fonógrafo ambulante focuses on a traveling phonograph, brought to an unnamed Andalusian village by well-educated science enthusiast and entrepreneur Restituto and his assistant Antero. Antero is soon revealed to be a young man originally from Aragon, in Northern Spain, who took up the job with Restituto solely with the intention of being able to travel around Spain and reunite with his beloved Araceli, who lives in the village and is due to get married, against her will, to the village's mayor. In the opening scenes, the phonograph threatens to disrupt the idyllic, traditional life of the village, with the villagers being simultaneously excited and concerned about Restituto's arrival. The work, therefore, initially echoes the tropes about rural Spaniards that were commonplace in press accounts of phonograph demonstrations, with recording technologies being portrayed as a disruptive force.

These tropes, however, are quickly subverted, and the phonograph shifts from threatening to benign when Antero's love for Araceli is revealed. It is indeed the phonograph's mobile nature which allows the couple to consolidate their relationship: firstly, because without it, Antero would have never been able to travel to Andalusia; secondly, because parading the device around the village gives the lovers an excuse to spend time together without their neighbors becoming too suspicious. In the context of the value system that dominates *zarzuela*, though, the role of the phonograph in bringing Araceli and Antero together is not necessarily transformative but rather restorative: Araceli and Antero are prototypical representatives of the Spanish *pueblo* coming from two different regions, and the phonograph becomes a positive force for them because it allows them to reunite. It does not follow from this, though, that Araceli, Antero, or most of their neighbors are invested in the broader ideas of modernization embodied by the phonograph. In fact, the only character thus enlightened is Restituto, who sees a connection between his own ideals of modernity and the increased connectivity between Spanish regions that the phonograph allows.[48]

El fonógrafo ambulante, though, does not solely focus on physical mobility: important in the play is also how the phonograph is able to move between social classes (from well-educated Restituto to the local mayor to the *pueblo* itself) and, more tentatively, how it might bring such social classes together. It becomes clear rather soon that Restituto is not part of the *pueblo*: his Spanish is more polished, and he repeatedly proclaims that he believes science is the only guarantee of progress for Spain, which causes some incomprehension and hilarity in the community. When he brings the phonograph to the *pueblo*, he does so from a position of privilege and never seriously mingles with the locals; however, he is portrayed as a sympathetic, well-intentioned character throughout, even if his being out of touch with the *pueblo* sometimes becomes the object of empathetic ridicule.[49] The mayor is portrayed in a less sympathetic way, as was often the case with *caciques* (local leaders) in *zarzuela*: indeed, he collaborates with high-up politicians to rig elections in their own benefit and achieve the majorities demanded by the *turno pacífico*, disregarding the interests of the *pueblo*. In the last scene of the play, however, after confusion has been cleared and Araceli and Antero have come together, all the characters are united in silently and admiringly listening to the phonograph.

We should not conclude from the preceding plot summary that Chapí and his librettist were trying to claim that the phonograph, and technology more generally, could crucially defeat social difference. *El fonógrafo ambulante*, like *zarzuela* more generally, does not purport social equality in the liberal sense; instead, it celebrates the virtues of an idealized, ultimately conservative *pueblo*—sometimes vis-à-vis those of the most privileged sectors of society.[50] In this sense, the last scene has something incongruous to it—as if Chapí and his librettist were poking fun at an implausible scene of temporary harmony between social classes before the phonograph's magic vanished and tensions inevitably began again. The scene therefore becomes an invitation to consider how the phonograph was accommodated, sometimes contradictorily, within the multiple interests in and visions of science upheld by different social classes and groups in Spain at the time—an approach I pursue in the next section.

Moving between classes: Phonographs in sociability spaces

Although the governments, the ruling classes, and the bourgeoisie were the most visible actors leading the development of science and the discourses around it in nineteenth-century Spain, other social groups and sectors (from the working classes to the army or the Catholic Church) also recognized science's transformative potential and even developed their own ways of understanding science and its role in the modernization of Spain.[51] It is therefore not surprising that practically all sectors of Spanish society welcomed the phonograph with considerable enthusiasm—from the aristocracy to working-class associations. The army also hosted demonstrations in its sociability spaces,[52] and in fact some phonograph demonstrators, as well as some *gabinete* owners later on, had a background in the engineering and technical echelons of the military.[53] Even the Catholic Church, which we might not typically associate with an enthusiasm for science, hosted phonograph demonstrations at some of its parishes and seminaries.[54]

Nevertheless, even the rather limited and fragmentary existing sources suggest that recording technologies hardly acted as an agent for social harmony in *fin-de-siècle* Spain, as half-heartedly suggested in the last scene of *El fonógrafo ambulante*. These demonstrations would take place in existing sociability spaces, public or semi-public, with a specific, if often fluid, class profile; most of these venues, implicitly or explicitly, welcomed individuals of certain social classes while excluding others more or less decidedly. The phonograph, by virtue of being present in all of these different spaces and conquering some new ones as time went by (for example, the ad hoc *salones* that proliferated in the middle and later part of the decades), made the frictions between them more visible. Studying the types of venues where phonographs were exhibited and demonstrated, as I will do in the pages that follow, can, therefore, deepen our understanding of how recording technologies were absorbed into Spain's social fabric; conversely, studying these venues from the point of view of phonograph exhibitions can add nuance to our understanding of sociability and scientific thinking in nineteenth-century Spain.

As happened in other countries throughout the Western world, Spanish sociability experienced significant growth in the nineteenth century starting in the 1830s, starting with the bourgeoisie and middle classes.[55] Two distinct models developed: the *ateneo*, which emphasized education and culture, and the *casino*, which favored leisure and recreational activities. This does not mean, however, that science and leisure were mutually exclusive: *ateneos* had spaces and occasions for more relaxed social interaction too, whereas *casinos* typically had a newspaper and magazine cabinet and often hosted scientific demonstrations.[56] Other types of sociability spaces included *círculos*, *clubes*, and *sociedades*. Generally speaking, these terms were rather loosely defined, and they could mean different things in different parts of Spain or different moments throughout the twentieth century, but most typically featured education and leisure as part of their activities. Sociability structures did not only exist for the benefit of its members: over time, many of them came to see themselves and be seen as a small-scale version of the liberal state—at a time in which the liberal state itself was in the process of being built.[57] For some, this meant putting their scientific know-how and interests to the service of society: for example, electricity was introduced and implemented in Spain mostly through the efforts of private societies with significant sociability elements.[58]

Bourgeois private clubs were defined not only by who their members were, but also by who was excluded from membership.[59] Exclusion could be purely economical, since associations typically required members to pay an admission fee upfront as well as annual or monthly subscriptions that might have discouraged those of modest means. But there were other ways in which clubs self-selected their members: membership often required an application process or the backing of existing members of the club, and clubs also organized around characteristics other than social class or location (for example, by type of profession or by political ideology). Partly as a response to these bourgeois associations, the working classes started to create their own sociability spaces toward the end of the nineteenth century. These were often reminiscent in spirit and name of their bourgeois equivalents (*ateneo obrero*, *casino de artesanos*), but also had their own defining characteristics.[60] While all or most of these centers were committed to educating the working classes and helping them form communities, their ideological foundations could vary considerably. Some of the centers were led by the workers themselves and organized under anarchist or socialist principles, while others had more reformistic intentions, and were led and organized by factory

owners, or by the Catholic Church.[61] The latter group was often under close su-
pervision from the authorities so as to avoid dissidence and rebellion.[62]

Both bourgeois and working-class sociability venues were frequent hosts of
phonograph exhibitions, and it can be presumed that admission to phonograph
demonstrations would be limited to existing club members. Therefore, when
such occasions were advertised in the local press, we must bear in mind that the
aim was not just to alert members who might not be aware of the event, but also
to showcase the club and their activities to those who were excluded.

Unsurprisingly, casinos—the most prominent typology of bourgeois private
clubs—proved to be very popular venues to host phonograph demonstrations.
Even though casinos had a strong recreational component, and were particularly well known as venues for gambling and fine dining, they were also keen to
promote scientific discoveries among their members, not only through talks,[63]
but also through the implementation of innovations, such as electric light, in
their own premises.[64] It is therefore unsurprising too that demonstrations at
casinos were typically promoted as eminently scientific, with the demonstration
itself often being carried out by a member of the casino with scientific or educational credentials (a "gentleman-scholar"), as was the case in Jerez in 1895 with
Gumersindo Fernández de la Rosa, a local engineer and also benefactor of the
local library.[65] The Marquis of Alta-Villa, visiting from Madrid, also presented
a phonograph in the casino of Guadalajara in 1898, as this would have likely
been the most appropriate venue in the city given his class background.[66] On
some occasions, casinos hosted itinerant funfair operators, as was the case with
Charles Kalb in Girona in 1890.[67]

Whereas casinos were indeed the flagship institution of Spanish bourgeois sociability, similar structures directed toward middle- and upper-class men also
hosted exhibitions, such as professional or business associations (the Círculo
Lebrero in Jerez,[68] the Círculo Mercantil in Madrid,[69] the Sociedad de Fomento
in Tarragona[70]), and so did artistic circles, which again would typically gather
together prominent locals with an interest in the arts or culture.[71] Some private
demonstrations taking place in aristocratic or bourgeois circles were mentioned
and discussed in the press in similar terms to those taking place in casinos and
other bourgeois sociability venues.[72] Interestingly, demonstrations in *ateneos*—
which were meant to be more focused on science than demonstrations in *casinos*
were—were likely scarcer: there is, indeed, only one documented instance at
the Ateneo de Madrid—in 1895 with Armando Hugens.[73] This suggests that
phonographs were not simply seen as merely scientific curiosities at the time: its
extended presence in broader centers of bourgeois sociability suggest that they
had a significant symbolic value too, becoming a living embodiment to its
members—as is the case in *El fonógrafo ambulante*—of the gradual modernization of the country and the perceived role of the middle and upper classes in it.

At the same time, the working classes also came to familiarize themselves with the phonograph—albeit in their own venues, and under slightly different discourses. There is no evidence that the working classes came to know phonographs belatedly with respect to their richer counterparts: in fact, one of the very first demonstrations of the Perfected Phonograph in Spain took place in September 1890 in the theater of the working class Barcelona neighborhood of Sant Andreu de Palomar, as a fundraiser for unemployed workers.[74] The announcement in the newspaper *La vanguardia* chose to highlight the fact that members of the audience would be able to experiment with the phonograph by themselves—suggesting that, apart from the fundraising component, the event was meant to be primarily about empirical science rather than about entertainment. Other notices about phonograph demonstrations taking place in working class sociability venues similarly put the stress on the scientific component, with an explanatory talk typically preceding or following the demonstration per se: this was the case of Armando Hugens at the Centro Instructivo del Obrero in Madrid[75] and local physics teacher Mariano Reimundo at the Círculo de Obreros in Salamanca, in which actually the first piece that its choir ever learned was reportedly recorded on a phonography by Reimundo himself.[76] A session at the Academia Obrera de Santo Tomás de Aquino in Valencia unusually put a focus on the spoken too, with workers reading poems and prose alternating with phonographs demonstrations.[77] The activity was organized by the Section for Religion and Morals of the Academia, suggesting that the reading and scientific session was intended, first and foremost, as morally healthy and virtuous entertainment, distinct from other forms of entertainment, such as commercial theater, regarded as damaging for workers—and perhaps distinct from other types of phonograph demonstrations aimed at the working classes that were starting to thrive at this time at the *salones fonográficos*, on which more later. Unlike demonstrations in casinos and other bourgeois sociability centers, events at working-class associations did not have the same aura of exclusivity: advertisements and reports focus instead on how these occasions encouraged a wide range of audiences to get acquainted with the phonograph.[78] This is consonant with the role that these working-class sociability centers played in late nineteenth-century Spain and with their commitment to improving the living conditions of the working classes through education and science.

However, sociability spaces in nineteenth-century Spain were not limited to private clubs subjected to a more-or-less stringent admission process. Other sociability models where membership was understood in a somewhat looser sense also became more popular throughout the century. Two of those models specifically frequently hosted phonograph exhibitions: cafés (and, secondarily, other hospitality structures such as hotels and *fondas*) and theaters, which were likely the most popular type of venue for such events. In theory, demonstrations at theaters and cafés were more inclusive than those taking place in casinos and *círculos*: potential audience members would need to be affluent enough to afford the price of a ticket or a drink, but they would not be subjected to any further membership requirements or demands for regular payments. This might have made such venues appealing to phonograph demonstrators, as they could reach larger, unfiltered audiences. A further advantage in the case of theaters, particularly in smaller provincial capitals and sizeable towns, was the fact that those tended to be centrally located, easily accessible, and well connected to the local press, so that advertisements and notices could be published.

Class considerations, nevertheless, still mattered considerably in cafés and theaters. Accounts of phonograph demonstrations in these spaces help us understand in some respects how fluid or otherwise class boundaries were there. In the provinces, theaters and, to a lesser extent, cafés, were prominent sociability centers for the bourgeoisie and the middle class, including well-off landowners and artisans.[79] It was also the middle classes—that is, those who could afford to travel for pleasure—who tended to visit hotels and *fondas*. We can therefore imagine that some of the demonstrations in these venues would be surrounded by an aura of prestige and exclusivity, and indeed some operators tried to enhance this prestige by inviting prominent locals to speak at the event. This is the case with a demonstration organized by Hugens at the Fonda de Oriente in Córdoba in 1894. Alongside Hugens, speakers at the event included the civil governor of the province, Mr. Ortiz Casado, the Count of Cárdenas, local journalists, and Gumersindo Fernández de la Rosa—who has been mentioned in this chapter as the organized of another demonstration in Jerez.[80]

However, while we might generally want to assume that demonstrations in theaters and cafés were attended mostly by the local bourgeoisie and middle classes, some sources suggest that at least a few of these events accommodated more diverse audiences.[81] An illustrative example that allows us an insight into the class tensions that such events might have elicited comes from a demonstration organized by Lorenzo Colís in June 1894 at the Teatro Coliseo in Logroño. An anonymous journalist, writing for the newspaper *La Rioja*, observed that the make-up of the audience was very different from the norm, and indicated that the *plateas* (boxes) were empty.[82] The *plateas* were normally rented for one year and would hence only be affordable to the wealthiest individuals and families in a town or city. The writer used sarcasm to suggest that this meant that the local bourgeoisie was disinterested in science ("Those less suspicious than myself would draw conclusions that would surely be far away from the truth: that is, that our most privileged classes are completely divorced from science and its multiple applications"[83]).

A third type of venue existed alongside private clubs, theaters, and cafés: venues that were built or refurbished with the primary purpose of exhibiting and demonstrating phonographs. Precisely because they did not make use of existing spaces, they were also the most fluid in terms of class and status. These ad hoc spaces can be divided into two types. The first included stalls set up on the occasion of local *fiestas* throughout Spain. Such venues were meant to be temporary and could be found in both urban and semi-urban or rural areas. It is likely that stalls at *fiestas* were during this time the most popular way for Spaniards living outside the provincial capitals to get acquainted with recording technologies. The second type of venue included *salones* or *salones recreativos*, such as that set up by Mister Pertierra in Madrid, which were intended to have some permanence in a specific urban space. The sources suggest that venues of this type became dominant in Madrid or Barcelona from 1895 onward, while the "public demonstration" model was still ubiquitous in the provinces.[84] While itinerant demonstrations still emphasized serious scientific dissemination,[85] the *salones* in the big cities were more decidedly oriented toward entertainment.

We should not assume, however, that *salones* and *fiesta* stalls were free from the class considerations that have been mentioned when discussing casinos, *círculos*, theaters, and cafés. Even though they were, in some ways, innovative structures, they were surely inspired by others fulfilling related purposes, and inserted themselves within broader, longer-running practices which would themselves have complex class dynamics. For example, phonograph stalls at *fiestas* would surely have had numerous resemblances with venues intended for the exhibition of magic lanterns, phantasmagoria, and other predominantly visual spectacles—not just in terms of their visual appearance, but also in terms of discourses and dynamics of class they partook of as a fundamental part of the *fiesta*. *Fiestas* themselves, although intended as a celebration of a particular city or town, were not free from class tensions, with the upper and middle classes organizing them through *comisiones* and otherwise taking up an important role in what we might consider the most symbolic, less recreational parts of the program, such as the religious ceremonies or the industrial exhibition of local industries that was often included in the celebrations. The more affluent social groups would also typically showcase their charitable activities by organizing a public giveaway of money, food, or clothes for the more disadvantaged inhabitants of the town.

Information on the pricing of phonograph auditions both at ad hoc stalls and at more permanent *salones fonográficos*, although far from comprehensive, allows us a glimpse into what sorts of listeners might have attended these places and what social classes they might have come from. Admission prices at these places could be as low as 5 *céntimos* and as high as 1 *peseta* (with 100 *céntimos* making up one *peseta*)—but not all *salones* or stalls offered exactly the same experience. Some businesses charged a flat fee and allowed customers to listen to as many cylinders as they wanted to.[86] Others charged their customers separately for each cylinder they listened to,[87] and sometimes sold "bundles" entitling customers to listen to several recordings at a slight discount.[88] It is likely that the latter approach was appealing to customers who could not afford the more costly flat fees; in any case, the variation in pricing approaches we find here suggests that phonograph operators were keen to consider a multiplicity of strategies to attract a range of social classes and therefore maximize their profits. Some also claimed explicitly in their publicity that they strived to keep their prices low so that people of all social classes could become acquainted with recording technologies.[89] Evidence indeed suggests that prices generally decreased as time went by and fell more decidedly under the 0.50 *pesetas* mark after 1897, particularly in Madrid in Barcelona. Advertisements from the provinces also show a gradual decrease, but sources are patchy and even non-existant for several of the provinces, and so it cannot be conclusively stated whether prices in the provinces were significantly higher than in the big cities or vice-versa.

An examination of the prices of *salones fonográficos* in the large cities indeed suggests that their offerings were among the most affordable in the culture and entertainment market, and hence financially accessible to a range of customers. For example, in Madrid, opera would be reserved for the working and upper middle classes, with a seat at the Teatro Real costing 15 *pesetas* in 1895, which amounted to several days' salary for a manual worker.[90] *Zarzuela* theaters were more affordable to a range of social classes—some of the most reputable (Español, Zarzuela, and Apolo), with seats costing between 4 and 6 *pesetas*, would still be predominantly attended by the middle classes,[91] but a range of less prestigious venues existed (Martín, Eslava, Novedades) where admission prices ranged from 15 to 75 *céntimos*.[92] Admission to a *salón fonográfico* was in line with the latter, costing between 10 and 75 *céntimos*. In some of the provinces, however, listening to a phonograph could be, comparatively, a rather expensive experience: in 1895 in Palma de Mallorca, an exhibitor was charging 1 *peseta* to listen to six cylinders,[93] whereas a seat at the local theater would have costed a mere 60 *céntimos*, and for 3 *pesetas* one would have access to a whole box.[94]

Despite the operators' efforts to open up to social classes, some of them also tried to signal in various ways that they adhered to bourgeois respectability and decency. When Colís installed his phonograph in Haro in September 1894, in an ad hoc stall at the Plaza de la Paz, the local newspaper hoped that the venue would soon become a meeting point for "good society";[95] similarly, a phonograph installed in 1898 at a bakery in Alicante was advertised as "the meeting point for the best families."[96] In Badajoz, a newspaper claimed that a local phonograph installation was very well attended by "young ladies" (*las señoritas*) suggests that the demonstrator wanted to give the event an air of respectability—as young women of certain social classes would have been unlikely to associate with venues regarded as not respectable.[97] A Mister Arenas, who installed in

1897 a phonograph in the Navas de Tolosa street in Pamplona, was reported in the newspaper to be very keen to ensure that "everything happens in agreement with decency" (*conforme a la decencia*) in his establishment.[98]

The article does not make clear what Mister Arenas meant exactly. It could be, for example, that he vowed not to make or play back cylinders containing obscene or inappropriate material: while there is no evidence that such cylinders were used in itinerant demonstrations or in *salones fonográficos*, a few *gabinetes* cylinders have survived containing jokes of a scatological nature, so it makes sense to think that such recordings would have had an audience earlier on. Mister Arenas might have also been aware that spectacles such as the phonograph gave unmarried couples opportunities to meet unsupervised (as is the case in *El fonógrafo ambulante*), and he might be keen to reassure local families that he would not allow this to happen in his venue. Similarly, it is difficult to ascertain whether operators like Mister Arenas were looking specifically at excluding certain sectors of the communities they were visiting, or whether they used respectability and decency as a further component of their publicity strategy—in a way so as encourage the working-class and lower-middle-class audiences who visited their shows to think of themselves as partaking in certain bourgeois values, and hence perhaps more likely to visit again. The latter might have been particularly the case for stationary establishments in Madrid or Barcelona, whose opening hours frequently extended until 11 p.m. or midnight,[99] hence running the risk of conflating recording technologies with other less respectable aspects of urban nightlife.

The phonographic demonstration as a genre

Despite social and geographical differences, evidence suggests that demonstrations held in public and semi-public spaces followed similar formats and conventions—what I term here the "phonographic demonstration." A demonstrator would play a series of recordings to the audience; sometimes, a scientific talk preceded or followed the event, and/or live music or spoken word was interspersed with the playing back of the recordings (of which more later). *Salones fonográficos* and stalls at *fiestas* followed slightly different conventions: not all of them relied on talks and live performances, but they instead encouraged spectators to navigate the phonograph themselves on their own or as a small group. However, the publicity and the general set up of the *salón* would no doubt have influenced the listening experience to a certain point. A 1900 article in *Boletín fonográfico* about these so-called *fonógrafos automáticos* (automatic phonographs) claimed that these were well-decorated spaces with phonographs and stacks of cylinders strategically distributed so that audiences stayed for a long time exploring the materials that were available.[100] There were some commonalities between public demonstrations and *salones*. Repertoires were practically identical; most importantly, as I will discuss in the rest of this section, both types of events still capitalized mostly on the phonograph's alleged ability to reproduce reality. At the same time, there is some evidence in both that new ways of listening to recorded music started developing around this time, with some audiences beginning to appreciate recordings as aesthetic objects rather than merely scientific curiosities, however.

The strongest evidence we have that audiences were mostly attracted by the phonograph's ability to reproduce reality is the fact that fidelity tests were commonplace in demonstrations. These are indeed to be found in other contexts up to the 1920s,[101] but in Spain they took a distinct form: normally, the operator would invite a local musician or speaker to play, sing, or speak in front of the phonograph; the resulting sound was recorded and then immediately played back to the audience. Examples are very numerous and come from demonstrations held throughout Spain.[102] In one of the first demonstrations of the Perfected Phonograph in Madrid, Mr. Sears and Mr. Warring invited the Count Aguilar de Inestrillas, a reputed army officer, and instructed him to record a voice of command which the phonograph allegedly played back to perfection.[103] Another event held later the same year in Madrid involved an anonymous soldier playing back a cornet solo.[104] Lorenzo Colís also played the cornet in his demonstrations to be played back immediately afterward, and also involved local singers for this purpose.[105] Luis Casares, a mechanic from Granada who went on to open his own *gabinete* and then to work for Pathé in France, recorded a tenor and a bass from the local cathedral in one of the demonstrations he organized.[106]

I would like to pause on two traits we systematically see in these demonstrations. Firstly, demonstrations, almost without exception, always used local singers and personalities for fidelity-testing purposes. Demonstrators could have recorded themselves or could have brought an assistant for this purpose; after all, this would have surely minimized the hassle of contacting prominent locals when arriving to a new town and securing their help. Secondly, recordings were always made in front of the audience and likely discarded afterward. Operators could instead have recorded their local contacts prior to the session, perhaps in the comfort of his or her own home. To record a local known to the audiences, and to do so in public, though, would have been necessary to make the fidelity test conclusive to audiences. Audience members would be familiar with the voice (or instrument) of the individuals being recorded, and they would have also seen them being recorded on stage, so they would be unlikely to suspect that the operator had somehow manipulated the recording or lied to them about who was in it.

Fidelity tests often got considerable attention in press accounts of the phonographic sessions: even though such accounts tend to be brief, fidelity tests were normally discussed in more detail than other parts of the event, suggesting that the scientific capabilities of the phonograph were still pretty much at the forefront during these demonstrations. Moreover, comparatively few notices and advertisements indicate which performers, composers, or pieces were recorded and played back. Many of the notices and advertisements contain simply broad indications of the sorts of musical genres (mostly opera, *zarzuela*, flamenco, some traditional music such as *jotas*, and *cuplés*) and types of spoken word recordings (speeches, short stories, jokes, dialogues, even prayers[107]) the audience could expect to hear, but it is much less common to find the titles of the pieces and names of performers. This would provide further confirmation that what drew audiences to phonograph demonstrations was not so much the music that they could listen to, but rather the possibility of listening to recorded sound itself and to examine the phonograph with their own eyes and ears. Similarly, there are few indications that operators were concerned with actively selecting repertoire (with the exception of Pertierra, as will be discussed subsequently). When they were, though, they chose specific pieces for their ability to showcase the capabilities of the phonograph, rather than for their intrinsic musical value. These were normally pieces with orchestral and polyphonic textures that would have been seen as more challenging to record at this time; examples include the overture to Rossini's *Guillaume Tell* in 1894 in Logroño,[108] an unnamed opera quartet at Mr. Warring and Mr. Sears' demonstration in Madrid,[109] and a local instrumental sextet recorded by Ramón Lenguasco in Almería in 1894.[110] None of these recordings—as is the case with all others from the demonstrations—have survived, so it is not possible to ascertain the extent to which these experiments were successful. On the one hand, with phonographs being less efficient at recording groups of instruments at this time than they were in recording voices (either unaccompanied or with piano accompaniment), we can presume that fidelity and quality would be disappointing from a present-day perspective. The very fact that several instruments could be recorded together, however, might have been enough to get approval from audiences who were in the process of getting acquainted with recording technologies—as suggested by a review of the Logroño event which claimed that "all instruments are perceived with the utmost clarity."[111] The widespread recording of the human voice might have pursued similar aims. The main reason why vocal recordings predominated in most of these demonstrations likely had to do with the fact that the voice was the easiest instrument to record with the technology available. A further factor, though, might have been the notion that voices were (and are) typically thought of as unique to each individual and hence more useful in proving the phonograph's fidelity capabilities: most audience members would likely not have been able to tell whether the violin they heard in a recording was their neighbor's or someone else's. However, almost everyone would immediately be able to recognize specific voices, and so the phonograph's ability to allow this type of recognition to happen might have been regarded as further proof of the device's scientific significance.

Compared to itinerant demonstrations, though, *salones* were more likely to indicate in their advertisements the names and titles of specific works, composers, and performers that audiences could listen to in their venues. Pertierra was one of the most active in this regard: he periodically added new cylinders to the collection in his *salón*, and advertised any new additions in the press.[112] Pertierra's repertoire included mostly selections from opera (*Faust, Dinorah, L'Africaine*) and *zarzuela* (*El gorro frigio, Certamen nacional, La verbena de la Paloma*) sung by performers active on the Madrid stages (Máximo Scaramella, Julia Segovia, Ángela Llanos, Emilio Mesejo, Joaquina Pino, Loreto Prado, Lucrecia Arana), as well as traditional music (*jotas* by the Aragonese singer Royo del Rabal), flamenco (by a Mister Revuelta), wind bands, spoken theater (*Inés del alma mía*), and short stories by Mister Domínguez—a comedian who would then go on to record short stories and jokes for the *gabinetes*, becoming the most successful recorded spoken word artist of the country for a while.

There is no other demonstrator or *salón* for which we have such a detailed account of the sorts of sounds their customers could expect to hear, and this allows us to draw some tentative conclusions as to how listeners might have received Pertierra's offerings. Itinerant demonstrators probably did not need to offer a wide range of repertoire to ensure customer interest: in many cases, the novelty of hearing a phonograph would be enough to attract a crowd and, with demonstrations and stalls being open for a few days or weeks at most, there was little risk that listeners would tire of listening to the same recordings again and again. Pertierra, on the other hand, had to play a longer game in Madrid, and it is likely that he felt the need to offer something else to make sure that customers who were sufficiently reassured of the capabilities of the phonograph kept coming back; an effective way of doing this might have been to renew his offerings periodically. On the other hand, we do not have any testimonies from Pertierra's audiences that allow us to calibrate how they made sense of the experience: whereas it might be that some kept going to the *salón* to reassure themselves that the phonograph could indeed handle different types of human voices in different genres, we might also entertain the possibility that a different way of thinking about recordings started to take shape at this point, with customers going back to Pertierra not so much to admire the phonograph's technical capabilities, but to enjoy different varieties of music and spoken word—some of which would be known to them from their live music experience, some of them not, as would then become more generalized in the era of the *gabinetes*.

Another argument in favor of the idea that some audiences and operators had already started to think, however inconsistently, of recordings as aesthetic objects and as commodities is the fact that the repertoire favored by Pertierra (and by other demonstrators that we know of[113]) coincides with the offerings recorded by the *gabinetes*: predominantly opera and *zarzuela*, and, secondarily, flamenco, traditional music, the spoken word, and wind bands, with instrumental solos being even more limited in number and art song being practically nonexistent. These choices coincide with the repertoires that were fashionable in Spain—particularly in urban settings—during the 1890s across different social classes, from the aristocracy and bourgeoisie (opera) to the working classes (*zarzuela*). There are a few instances of more marginal repertoires, which suggests that, in a small number of cases, recordings might have been seen as a medium for audiences to get acquainted with unfamiliar sounds. For example, an alleged mechanic called Mister Johnsson (*sic*) offered demonstrations in early 1895 in Barcelona with religious music sung by the *escolanía* (boys' choir) of Merced,[114] or a session at the Unión Católica in Madrid in 1895 where compositions from members of the association were recorded.[115] Interestingly, Lucio García Leal played mostly French music in demonstrating the phonograph Lioret, which suggests that the cylinders he had were pre-recorded and imported from France: the newspaper reporting on the event indicated that the only Spanish-language cylinder was a recording of a *jota*.[116]

Pertierra's repertoire lists suggest that he was successful in hiring some of the most prominent names of 1890s *zarzuela*, thus providing a further avenue for aesthetic rather than scientific listening. Customers would not come to his establishment necessarily to admire the technology, but rather to hear famous singers they knew from the stages or they had heard about, such as Loreto Prado, Lucrecia Arana, and Joaquina Pino. Here, the connection to the *gabinetes* is not as straightforward as with the repertoire, since the *gabinetes* employed, to a great extent, lesser-known singers: indeed, Prado and Pino never recorded for them, whereas Arana did so only to a limited extent. Potential reasons why famous singers generally did not record for *gabinetes* will be discussed in Chapter 6, but it is likely that, while celebrity singers could be persuaded to record on a one-off basis for Pertierra or other demonstrators,[117] it is more difficult that they would agree to submit themselves to the more sustained working processes of the *gabinetes*, where they would often be asked to produce several copies of the same recording in one session, as will be discussed in the subsequent chapters. We should also not dismiss the possibility that some demonstrators lied about the identities of the performers in their recordings: a phonograph demonstration at the Sociedad de Fomento in Girona advertised recordings of *La favorita* by Julián Gayarre (with piano accompaniment),[118] but there is no other evidence that Gayarre ever recorded (commercially or otherwise),[119] so the mention of his name might have been intended simply to lure audiences into the session. In any case, given that none of these recordings have survived, it is not possible to arrive at more solid conclusions.

Conclusion

It would be exaggerated or inaccurate to claim that traveling phonographs revolutionized Spanish life during the 1890s and early 1900s, given that recording technologies still were, to an overwhelming majority of Spaniards, a rarity to be admired for scientific and technological reasons rather than as an everyday object. Nevertheless, the itinerant demonstrations and stalls and, later on, the *salones* allowed a considerable part of the Spanish population to get acquainted with recording technologies, with demonstrations taking place in most provinces and across all social classes. The more widespread presence of the phonograph in all of these spaces and—perhaps most relevantly—traveling from one to the other allowed the imaginaries of recording technologies to expand: phonographs were indeed saluted as an embodiment of mobility and therefore of modernization, in line with discourses about national regeneration active in Spain at the time. Nevertheless, a closer examination of the sources reveals that the arrival of the phonograph was not unproblematic: it was not always seamlessly integrated into such discourses, particularly those which idealized connection between the regions and/or tacitly excluded some social classes from modernization and science processes to the detriment of others. It cannot be said that the phonograph or its demonstrators challenged these discourses, but they certainly contributed, from our perspective, to understanding some of the inconsistencies in them. They also gave rise to some new ways of thinking about recording technologies that came to be forgotten or minimized in future stages: for example, the idea of mobility being a key part of recording technologies became obsolete in the era of the *gabinetes* and beyond, where the phonograph became stationary and firmly ensconced in the home. Compared to the initial, very sporadic appearances of the phonograph in Spain in the period 1877–1888, class made a more decided appearance as well in terms of who got to listen to phonographs and under which circumstances; this would indeed pave the way for phonographs becoming a middle- or upper-class accessory in the *gabinetes* era. The utopian vision presented at the end of *El fonógrafo ambulante* does not seem to have materialized in 1890s Spain.

But, apart from geographical and class considerations, what we might find most unusual about the experiences of Spaniards listening to traveling phonographs in the 1890s is the mechanics of the actual phonograph-listening: collectively or at least in a public space, surrounded by others with whom to comment on the experience, rather than in the privacy of one's home; to reassure oneself that Edison's marvelous invention worked, rather than to enjoy the music of one's favorite composers or the voice of specific singers. But some Spaniards, at least in some contexts, might have started to look beyond the technology and the fascination with the reproduction of reality to pay attention at what actually was in the cylinders; nevertheless, it would be the *gabinetes* who more decidedly encouraged aesthetic listening by inventing the recording as a cultural artifact for the Spanish context.

3

Inventing the recording

Gabinetes fonográficos and early commercial phonography in Madrid, 1896–1905

Edison's National Phonograph Company, founded in January 1896, opened the door for the phonograph to be transformed from scientific curiosity to purveyor of domestic entertainment. Within two years, three new versions of Edison's invention were launched into the US market: the Spring Motor Phonograph in 1896, the Edison Home Phonograph mere months later, and finally the Edison Standard Phonograph in 1898[1]—all of them easier to operate and more affordable, at least to the middle classes, than their predecessors had been. Alongside new versions of the phonograph, it was also in these years that Edison developed the wax cylinder as its recording support. Technological and commercial innovation aside, Edison's inventions from these years contributed crucially, too, to the development of cultural and aesthetic concepts that still govern, to various degrees, our way of relating to music, such as the concept of solitary listening or—perhaps most influentially—the notion of recorded sound as a commodity. Such concepts did not develop in a linear, straightforward way, as a product of technological determinism; instead, they resulted from negotiations, frictions, and contradictions, sometimes of global reach, sometimes closely connected to specific national, regional or local practices and debates. In the following three chapters, I focus on specific geographical areas in Spain (Madrid in Chapter 3, Barcelona in Chapter 4, the Spanish provinces in Chapter 5) to show how the commercial recording was born at the hands of local companies called *gabinetes fonográficos*, developing under different pressures and in different circumstances in different parts of the country while still showing certain national traits. As I will discuss in detail, influence from the national and local context in the shaping of recording technologies and their uses took a variety of forms. Sometimes such influences were material and tangible, as is the case with urban geography shaping the nascent recording industry in very different ways in Madrid and Barcelona, and sometimes they operated in the realm of the abstract, with nation-wide debates and ideas about science, technology, and modernization influencing how individuals conceived of the phonograph and recordings. Due to the level of activity of the Spanish *gabinetes* and the multiple surviving sources they have left us compared to other countries—for example, with

Inventing the Recording. Eva Moreda Rodríguez, Oxford University Press. © Oxford University Press 2021.
DOI: 10.1093/oso/9780197552063.003.0004

neighboring Portugal—Spain provides us with a unique case study to examine how these local and national synergies influenced and were in turn influenced by the advent of sound technologies and their absorption into the fabric of society. At the same time, the *gabinetes* did not operate in isolation, but instead were part of transnational networks allowing the circulation of new products, but also of new discourses and ideas that were reflected in different ways.

In the period under study, about forty *gabinetes fonográficos* existed throughout the country, operating independently from one another. They sold phonographs, graphophones, and gramophones imported from France or the United States; there were not, at the time, any phonograph or gramophone factories in Spain, although Spanish entrepreneurs did engage in a few manufacturing experiments, some of which will be discussed over the course of the following three chapters. *Gabinetes* also sold accessories for talking machines, such as diaphragms and headphones, some of them imported and others developed locally by the *gabinetes* themselves. Most importantly, though, they produced and sold wax cylinder recordings, employing mostly local opera and *zarzuela* singers. Some *gabinetes* also imported and sold impressed wax cylinders from France and the United States, but it was their locally produced cylinders that they advertised the most and what helped them build an identity and position themselves within national debates. Whereas in other countries, such as India, local businesses depended on multinationals to process cylinders after recording (in the case of India, with Pathé),[2] the Spanish industry was self-sufficient in this respect, which allowed *gabinetes* to develop in a remarkably autonomous way: evidence indeed suggests that the *gabinetes* operated almost independently from Edison and his companies. It is likely that Edison officials were aware to some extent of the *gabinetes* industry, as *gabinetes* bought phonographs and blank cylinders from their companies, but, other than that, interactions between both industries seem to have been limited or non-existent. For example, while Edison and other early major players such as Columbia were involved in complex feuds with smaller companies in the United States, acquiring them or suing them for patent infringement,[3] the Edison Papers contain no evidence that they were bothered in the slightest by local developments in Spain. One reason for this might be that the Spanish market was, at the time, a comparatively small and remote one, and so it would not have initially posed much of a threat to Edison and his companies; this laissez-faire attitude, together with the local particularities that will be discussed in the following three chapters, was likely a contributing factor for the Spanish early recording industry to develop its unique identity.

Not all of the *gabinetes* discussed in this chapter and the following two were active for the entirety of the period 1896–1905. In fact, many of them were extremely short-lived, as is obvious from Table 3.1. Books and records from *gabinetes* have not survived, wax cylinders are not dated, and catalogues were

Table 3.1 *Gabinetes fonográficos* in Madrid, 1896–1905

Name	Dates active[a]	Key individuals	Own recordings?	Other activities	Other
Álvaro Ureña	1899–1903 or 1904	Álvaro Ureña	Yes	None	Became a Gramophone representative in 1904
Antonio G. Escobar El Graphos	1900–1904 (?)	Antonio G. Escobar	Yes	Photography, cinema	El Graphos appears as a recording seller in the *Anuario del Comercio* 1904; however, none of its press advertisements mention phonographs or recordings after 1901 (only photography)
Bazar de la Unión	?	?	Unclear	Department store	It is likely that this department store sold phonographs and gramophones, but no evidence of own recordings
Comisariato Internacional Comercio	?	?	Yes	Building society	
Fono-Reyna. Sociedad anónima fonográfica	1900–1904	Perhaps baritone Juan Reyna?	Yes	None	
Guido Giarretta	1900–1903	Guido Giarretta	Unclear	Bicycles	
Hugens y Acosta / Sociedad Fonográfica Española	1896–1905	Armando Hugens, Sebastián Acosta	Yes	None	Traded as Hugens y Acosta first, then Sociedad Fonográfica Española (presumably after Acosta left the partnership)

Establishment	Dates active[a]	Owner(s)	Gabinete fonográfico	Other activities	Notes
J. Oliva	?	J. Oliva	Unclear	Scientific equipment, optician, electricity	
José Navarro / La primitiva fonográfica	1900–1904	José Navarro Ladrón de Guevara	Yes	Clockmaking	Operated as José Navarro from 1900 and as La primitiva fonográfica in 1904–1906 (selling imported recordings). Then active until ca. 1910 as Fábrica de Grafófonos.
Julián Solá	1900–1901	Julián Solá	Unclear	Electricity	
La fonográfica madrileña	1902–1905	Atanasio Palacio Valdés	Yes	None	Active until at least 1909 as a reseller of Odeon and Fonotipia
Obdulio B. Villasante	1899–1901	Obdulio B. Villasante	Yes	Optician, electricity	
Viuda de Aramburo	1897–1900	Juan González González, Teresa Quesada	Yes	Sale of scientific equipment; electricity; photography	

[a] "Dates active" refers to the dates for which we know (through press advertisements and articles, or through inclusion in industry yearbooks) that a particular establishment was operating as a *gabinete fonográfico*, that is, selling phonographs and/or recordings. The establishment may have been active selling other types of products or services before or after that.

scarce (some of them will be discussed in due course), so the most reliable sources to establish the chronologies of individual *gabinetes* are notices and advertisements in the press, as well as commerce yearbooks such as *Anuario Riera* and *Anuario del Comercio*, typically published annually at the beginning of the year and providing a directory of the businesses operating at that point in a given territory. But there are issues with both types of sources—which are illustrative more generally of the patchy, haphazard nature of the evidence available to reconstruct details of the *gabinetes* era. While some *gabinetes*, such as Hugens y Acosta in Madrid or the Valencia *gabinetes*, actively and systematically advertised themselves in the local press, others, such as El fonógrafo in Barcelona, chose not to do so: we know that El fonógrafo existed because ten of its cylinders have survived, but we do not know when exactly it was active. Other *gabinetes* advertised rather aggressively for a few months after they opened, but disappear from printed sources after that. In such cases, it is not normally possible to tell whether the *gabinete* effectively closed down after that, or whether they simply stopped advertising in the press but continued trading. Another issue is that most *gabinetes* did not occupy themselves exclusively with phonographs and recordings. Instead, they typically operated as a side line to existing businesses, normally in the fields of applied science and technology (pharmacies, drug stores, opticians, electricians, photographers), which is fully consonant with the way in which domestic recording technologies were introduced in markets elsewhere in the world, in countries as different as Portugal and India.[4] In a trade yearbook for a given year, an establishment might have been listed as providing products and services in the fields of both photography and sound technologies; the next year, the latter might have been dropped, while the establishment was still listed under the photography rubric. The most straightforward conclusion to this is that the establishment had decided to stop trading in sound technologies but kept selling photo cameras and film. We must bear in mind, however, that such yearbooks were not always completely accurate, and that some businesses might not have provided full details of all of their activities, particularly those they only pursued on a part-time basis.

The short lives of many *gabinetes*, as well as the fact that most of them were side lines to existing businesses, speak of a rapidly moving climate in which professionals with reasonably relevant expertise identified a business opportunity in sound technologies and pursued it, to varying levels of success. Their motivations for doing so were not unanimous, as will be discussed in the following three chapters: some were genuinely interested in introducing technological innovation and contributing to modernizing Spain; a minority might have been more interested in the music itself; finally, others might have been motivated primarily by financial gain. One reason why the industry moved so rapidly in its early years likely had to do with the fact that barriers to entry were low: a phonograph, a few blank wax cylinders, a piano, and a reasonably sized room, as well as contacts with local musicians, would have sufficed for a *gabinete* to start operation. Apart from singers, who could be hired on an ad hoc basis, *gabinetes* did not need much personnel either: some of them were essentially a one-person operation, with the same individual supervising recording sessions and looking after sales, marketing, accounts, and other activities. A small number of *gabinetes* employed an artistic director, normally a musician by profession, who, on a part-time basis, would typically select and approach singers, supervise recording sessions, and perhaps accompany them on the piano. With many *gabinetes* operating part-time or over a limited time span, and with the Spanish economy being smaller and less articulated than the US one, the early Spanish recording industry was a comparatively small, fragmented one, relying to a great extent on artisanal processes rather than on mass production.

The early days of the recording industry, in Spain and everywhere, were certainly heavily shaped by the opportunities and limitations posed by the technologies in use. This determined, for example, what sorts of repertoires, instruments, and voices could be recorded, how fast recordings could be made, and how the resulting recordings could be sold and circulated.[5] Of course, such limitations also operated in the *gabinetes* industry; there was one in particular that became especially important in shaping their activities in multiple ways, as will be discussed in the subsequent chapters up until the end of this book: from their choice of singers to their focus on artisanal rather than mass production, to the ontologies of recording they spread among their customers. This limitation was none other than the fact that Spanish *gabinetes* did not manage to develop a reliable, industrially viable method of impressing multiple copies of the same cylinder. This had indeed been a challenge elsewhere in the world: back in the United States, Bettini and Edison were able to duplicate cylinders by using a pantograph, but this had significant limitations: copies were rarely the same quality and the original, and only a maximum of about 150 copies could be produced in total. Although no conclusive evidence exists that pantographs were used in Spain, articles in *El cardo* suggests that Viuda de Aramburo and other *gabinetes* might have used an unnamed machine to duplicate cylinders.[6] A further article in *Boletín fonográfico* claimed that Hugens y Acosta "duplicated" cylinders by recording the same performance in two or three phonographs at the same time.[7] As will be discussed in more detail in the course of this chapter, *Boletín fonográfico* and other *gabinetes* were normally critical of such duplications, arguing that they were of very bad quality. Some *gabinetes* and independent inventors also tried to develop their own techniques and technologies for cylinder duplication, although there is no evidence that they were ever implemented commercially.[8] We should therefore operate under the assumption that most Spanish wax cylinders from this era were therefore unique or quasi-unique—a one-off record of a particular performance—with this limitation shaping the operations and the culture of the industry in significant ways.

The preceding considerations about *gabinetes* apply to all of Spain. Location, however, influenced other aspects of the industry in key respects, so that it makes sense to discuss *gabinetes* in Madrid, Barcelona, Valencia, and the rest of the country separately. Location not only shaped the *gabinetes'* activities in material, tangible ways, determining, for example, the sorts of customers and musicians they had access to: it also influenced them on a more abstract level, as discourses around science, technology, and *regeneracionismo* had different undertones in different parts of the country.

Researching *gabinetes fonográficos* in Madrid, 1896–1905

Among Spanish cities, Madrid counted the highest number of *gabinetes*; some of these were among the longest-lived in Spain, and probably among the highest-producing too, even though it is not easy to ascertain how many wax cylinders were manufactured and sold at this time in Spain.[9] This is, to an extent, to be expected, given that Madrid was the most populated Spanish city at the time (although with Barcelona a very close second) and also the capital in a highly centralized state, making it the center of many of the discourses concerning science, technology, and modernization.

Table 3.1 offers an overview of *gabinetes* active in Madrid during this period; similar tables will be provided in Chapters 4 and 5 for Barcelona and the provinces.

In contrast with *gabinetes* in other cities, the Madrid *gabinetes* were also more successful in articulating and disseminating a discourse surrounding recording technologies, recordings, and their consumption in the home that closely resembled some of the tenets of *regeneracionismo* concerning science, technology, progress, modernization, and national identity that have been discussed in the introduction. This discourse was certainly not disconnected from developments elsewhere in the world: the *gabinetes* were part of international trade networks, most notably with the United States and France, from which they imported some of the products they sold, and it is therefore to be expected that they would import some of the discourses that were developing in those countries too. This was facilitated by the fact that the *gabinetes* and their foreign counterparts spoke, to an extent, the same language: as was the case elsewhere (most notably in the United States),[10] in Spain it was primarily technology and applied science professionals who led the development of the early recording industry, and not musicians or music specialists, which could have resulted in a very different outlook. The Madrid *gabinetes* articulated these discourses through their advertisements, debates, and other interventions in the press, through the specialized publication *El cardo* and through the physical recordings themselves.

The *gabinetes'* discourse can therefore be regarded to pursue two aims. On the one hand, it aimed at developing the notion of the recording as a commodity, which is consonant with what had been happening elsewhere in the world, particularly in more industrialized nations like the United States, Britain, France, and Germany, since the early 1890s. Before the first *gabinete* opened in 1896, Spaniards had attended collective phonograph demonstrations wanting to convince themselves that sound reproduction was indeed feasible, but now the challenge for the *gabinetes* was to persuade at least some of those Spaniards—the most affluent or technologically minded—that the phonograph belonged in the household, and that it could be a source of domestic pleasure and aesthetic enjoyment rather than just a scientific curiosity. The second aim of the *gabinetes'* discourse was directed toward their local and national spheres, rather than the global: the *gabinetes* strived to connect the new products and practices emerging around recording technologies to *regeneracionista* aims, and to position themselves as part of the *regeneracionista* class. The reasons why the Madrid *gabinetes* found in *regeneracionismo* a compelling framework to situate their activities go back to before the advent of sound technologies in Spain: indeed, small business owners and liberal professionals in the field of practical and applied science and technology, like most of the *gabinete* owners were, had been for decades an important driver of the introduction of new scientific discoveries and technologies in Spain—even though their contribution has often been forgotten.[11] Although these professionals rarely conducted pure research like those employed at universities and research institutions did, they usually had some scientific training and they also tended to be aware of scientific and technological developments in their field abroad, which they often imported and adapted to the Spanish market. It is therefore understandable that at least some of them identified with the broader *regeneracionista* outlook.

We might ask ourselves why the Madrid *gabinetes* were able to articulate a discourse on themselves and their products where pioneers of phonography elsewhere in Spain failed. A reason might be that, being in the capital, they were also close to the center of *regeneracionista* debates and to the institutions where some of those debates took place. But we might argue that Barcelona *gabinetes* too were close to the equivalent Catalan institutions, which were at the same time engaging in debates about national identity and science, and yet did not manage to fully articulate a similar discourse. An answer to that might be that the individuals involved in each of these cities were different in personality and nature: indeed, in Madrid, a few very active *gabinete* owners were instrumental in articulating this discourse, and their energy and projects might have galvanized others. In Barcelona, on the other hand, leadership in the recording industry was more diffuse at this stage, as will be discussed in Chapter 4. It is noteworthy, too, that the most influential *gabinete* owners in Madrid were relative newcomers or outsiders to applied science: whereas most *gabinete* owners had worked as pharmaceutics, opticians, or scientific equipment manufacturers for most of their lives, this was not the case with Armando Hugens, Álvaro Ureña, and the Marquis of Alta-Villa, whose profiles I will discuss in more detail over the next pages. It is plausible that the very fact of being newcomers encouraged them to develop discourses which not only helped them sell their innovative products, but also assisted them in building a professional identity and winning the trust of their new colleagues in the nascent recording industry.

French-born Armando Hugens, one of the partners at Sociedad Fonográfica Española Hugens y Acosta,[12] was particularly prominent and visible among Madrid *gabinete* owners. It is unclear what Hugens's career was in France and when exactly he settled in Spain, but his first documented forays into the world of recording technologies in Spain date from 1893, when he translated from French into Spanish a pamphlet on the phonograph by Mathieu Villon.[13] For the next few years after that, he toured the country offering phonographic demonstrations (presumably for a fee) in several Spanish cities in Andalusia and the Balearic islands.[14] The very first advertisement of a *gabinete fonográfico* published in the Spanish press also came from Hugens, on March 15, 1897 in *El Globo*.[15] There is evidence that Hugens was operating an establishment of some sort in Madrid in December 1896 where customers could record their own or their relatives' voices in a phonograph (which therefore ties in with earlier ideas about the phonograph being used to preserve voices after death), although it is not clear whether he produced and sold his own recordings at this stage.[16]

Hugens's partner in the *gabinete* was a Spaniard called Sebastián Acosta. Not much is known about him: indeed, most press coverage about the *gabinete* focuses on Hugens, and suggests that it was Hugens who was in charge of technical and artistic aspects of the business as well as acting as its spokesperson, whereas Acosta focused on accounting and management. Acosta then left the business around 1901. Both Hugens and Acosta fulfilled, in different ways, what we might call the role of *regeneracionista* entrepreneur, committed to developing projects in his or her area and expertise and doing so not only for financial gain but also to help the regeneration of the country.[17] This is most obvious in an article about the *gabinete* printed in the newspaper *El liberal*: the writer of the article praised Hugens's scientific background and his good artistic taste, and also wrote admiringly about the fact that Acosta, having secured a life pension from the Spanish government after his civil servant job in Cuba disappeared with the loss of the colony, chose not to lead an idle life thereafter, but instead "strived to make the Spanish industry grow."[18] This was high praise: in the *regeneracionista* imagination, civil servants were often portrayed as conformist, lazy, incompetent, and excessively attached to bureaucracy,[19] so the fact that Acosta had decided to take a more entrepreneurial path in life was regarded as remarkable.

Álvaro Ureña, who in 1899 founded a *gabinete* of the same name in Madrid, was another pioneer active in shaping discourses around phonography. He was unusual among early *gabinete* owners in that his background was in the army, and during the 1880s and early 1890s he was an infantry sergeant and then an *escribiente* (army clerk).[20] In the mid-1890s, he took leave from the army to pursue his interest in technology and electricity,[21] and in 1895 he worked on the re-electrification of Alcalá de Henares, a town near Madrid.[22] He subsequently opened an electricity business in Madrid itself where he sold a variety of electrical devices and equipment. It was during these years that the scientific dissemination magazine *Madrid científico* praised him for developing a commercial and research strategy based purely on scientific knowledge and evidence, and not on sensationalism.[23] Like Acosta, Ureña was, as a former soldier, an unlikely proponent of *regeneracionista* aims, as *regeneracionistas* would often position themselves in opposition to what they considered to be the establishment, which often included the army and its allegedly incompetent leaders.[24]

Similarly unusual among those who were pushing the early recording industry was José Ramiro de la Puente y González Nandín, Marquis of Alta-Villa: a member of the nobility and a high-rank civil servant (*abogado del estado*), he accompanied Isabel II into exile between 1875 and 1882, working as her secretary. Back in Spain, he was one of the first to enthusiastically promote commercial phonography among the Madrilenian public: he never owned or ran a *gabinete*, but he started a phonography supplement in his very own magazine, *El cardo*, crucially contributing to articulating the discourse of the early recording industry in Spain, as will be discussed later. It is likely that Cilindrique, who signed numerous articles in *El cardo* commenting on the developments of the phonograph industry in Spain, was a pseudonym of Alta-Villa's. Although less active than the individuals discussed previously, we might also name, as pillars of the Madrid recording industry, José Navarro Ladrón de Guevara, originally a clockmaker who started off offering phonograph demonstrations in the Valencia-Alicante-Murcia area[25] and was active with his own *gabinete* in Madrid from 1900; and Atanasio Palacio Valdés (owner from 1902 of La fonográfica madrileña), who had a background in the military, politics, and librarianship.

Launch parties offered *gabinetes* an opportunity to stage and showcase in front of an audience the values that underpinned their activities, and the connections of those values to *regeneracionista* discourses. Importantly, some *gabinetes* threw launch parties after several months or even years of successful operation, which suggests that owners managed these events strategically, holding them at what they thought were the most convenient times for their business and not necessarily to mark their opening. Such events, as well as their press coverage, tended to emphasize the aura of luxury that surrounded *gabinetes*. Fono-Reyna Sociedad Artístico Fonográfica hired soprano Josefina Huguet and tenor Ángel Constantí for its official launch party in April 1899.[26] The soirée included opera but not *zarzuela*, as the former was regarded as more prestigious than the latter. Ureña, on the other hand, held a first launch party in July 1897 when he first started operation,[27] and another one in 1900, even though there is no indication that Ureña renovated his premises or otherwise introduced any changes that justified a re-inauguration. Whereas the 1897 party seems to have been comparatively modest, the 1900 one involved well-known soprano Elena Nieves, the marching band of the regiment of Corinola, and flamenco singers and dancers impressing cylinders on site. In reviewing the event, *La correspondencia militar* dedicated considerable attention to describing the luxurious premises of the *gabinete*:

> There cannot be more luxury, more art and more wealth in this establishment. The rooms are spacious, and from the ceiling hang elegant electric lamps; their golden and bronze-coloured filigree interlocks artistically and complements the power of the very modern voltaic arcs; phonographs of all sorts and mechanisms are scattered very tastefully around the place.[28]

Hugens y Acosta similarly organized its launch party after almost two years of operation, in December 1898. The newspaper *La época* described Hugens y Acosta's headquarters in the following terms:

> The installation is very interesting to see, and its taste and artistic elegance are first class. The shop, which opens directly to the street in front, is adorable. The audition room is a very charmingly decorated *bombonnière*. Cylinders are impressed in this room every day, and the next day they are transferred over to the shop window as the dish of the day offered to the phonographic gourmets.[29]

Other *gabinetes* put on events which, though not formally labeled as launch parties, allowed them to access the upper echelons of the population. Viuda de Aramburo was one of the first to do so, exhibiting in December 1897 one of its phonographs at a soirée at Círculo Mercantil—an association of industrialists. A Mr. Silva sang opera arias and an ensemble of plucked strings performing *jotas*.[30] In addition, the word "gabinete" has luxury connotations in itself: originally meaning "cabinet," throughout the nineteenth century its meaning shifted to denote a room within a *casino* where members could read newspapers or discuss literary topics—a highly exclusive space.[31] The *gabinetes'* luxurious spaces were reminiscent of those of *casinos* from the mid-nineteenth century onward,[32] which therefore reaffirmed *gabinetes* as exclusive venues accessible only to the middle and upper classes.

The *gabinetes'* focus on luxury might be seen as scarcely consonant with *regeneracionismo's* preoccupation with practical, useful science. Indeed, criticism of *despilfarro* (overspend), particularly among the nobility, church, and upper classes, was a key feature of the *regeneracionista* discourse.[33] Nevertheless, developing an aura of luxury around phonography was part of the necessary process to turn recorded music and the recording more specifically into a commodity, into something that could be bought, stored, and treasured, both revitalizing trade and stimulating cultural development in the process, to the exclusion of social sectors lacking the financial means or the technical ability to own and operate a phonograph. This is consonant with the formation of other technocultures in the nineteenth century around the world—from the United States to Colombia, where recording technologies were often presented as civilizing forces excluding certain populations by virtue of race, gender, or class.[34]

The Madrid *gabinetes*, therefore, likely faced a dilemma between the need to build an aura of luxury around their products which could help develop the notion that recorded sound was now a commodity, and the push to adapt their discourses to their cultural context. They found a solution in emphasizing the message that the luxury they stood for was practical, pragmatic, useful, and eminently based on scientific developments, rather than extravagant luxury—which suited *regeneracionista* ideals better. For example, reviews and descriptions of Ureña's *gabinete*, as well as his own publicity, suggest that he was keen to strike a balance between an aesthetically pleasing display, scientific rigor, and affordable prices.[35] Ureña termed the phonograph as "the most cultivated pastime known to man," but also emphasized that the phonograph was now an *affordable* luxury (affordable, at least, to the middle and upper classes): "In the past," he wrote, "to acquire a phonograph costed thousands of *reales*; nowadays it can be acquired for just 100 *pesetas*, with all of its accessories, including six impressed and two blank cylinders."[36] Ureña's reasoning was close to the *regeneracionista* ideal of scientific developments being imported into Spain to improve the standard of life of all its inhabitants. He followed a similar strategy with automobiles, which he also tried to introduce in Spain from the late nineteenth century onward.[37]

Among *gabinete* owners, Ureña was also one of the keenest to openly signal the steps that he was taking in making sure that the practical, affordably luxury he offered was integrated within the country's economy and society—in a way so as to advance *regeneracionista* aims, but never causing disruption and ensuring he was working together with institutions and not overly challenging them. Indeed, while the *regeneracionistas* initially positioned themselves as anti-establishment, or at least establishment-critical, by the time the *gabinetes* were active their ideas had spread enough that they were commonplace in many sectors of Spanish society, including government and administration, and so the *gabinetes* might have regarded direct confrontation as unadvisable. Ureña took part in 1901 in the Exposición madrileña de pequeñas industrias, aimed at encouraging small businesses, and obtained a prize for his cylinders.[38] When publicizing his phonographs and wax cylinders, he also emphasized his connections to the state institutions—for example, in 1902 he claimed that he had installed the electricity in the Congreso de los Diputados.[39] The newspaper La época also claimed that he had the intention to gift king Alfonso XIII a phonograph which was allegedly so technologically advanced that when exhibited at a high society party, attendants thought that well-known tenor Tamagno was singing live.[40] There is no record, however, of Ureña having done so; the story about party attendees might be similarly apocryphal and circulated by Ureña for publicity reasons, as it is fully in line with other publicity accounts that exaggerate the fidelity capabilities of the phonograph.

Gabinete Hugens y Acosta were also keen to promote their connections to the official institutions. In an advertisement in 1900 they claimed that they had been appointed to manage the phonographic archive of the Teatro Real and received exclusive permission to record the Teatro's singers.[41] If this were accurate, it would have been a significant event indeed, as no other Spanish theater or institution had started its own sound archive (and it would indeed be decades before more consistent sound-collecting policies emerged). Nevertheless, the claim does not appear in sources other than Hugens y Acosta's own publicity and there is no evidence that the Teatro Real had an archive at that point, suggesting that it might have simply been an attempt by the gabinete to further its legitimacy—or a project that never came to fruition. The claim that Hugens y Acosta had signed an "exclusive" contract with the Teatro Real is similarly dubious: the gabinete was, at the time, recording singers who were active at the Real and some of these singers might have individually signed exclusive contracts with the gabinete, but this does not mean that the contract extended to the Real as an institution; indeed, Real regulars such as Fidela Gardeta, Antonio Vidal, and Concha Dahlander regularly recorded for gabinetes other than Hugens y Acosta around this time. Similarly, in 1902 Hugens was advertising in the Guía de la coronación—a guide of Madrid issued for those planning to visit the capital for the coronation of Alfonso XIII. The advertisement presented Hugens as a committed patriot, claiming that he did not "shy away from sacrifice" to import the latest technological innovations[42]—therefore stimulating exterior trade and contributing to turn the capital into a hub of commercial activity for the benefit of those wealthy enough to visit for the coronation.

While some of the gabinetes did boast about their connections to public institutions, this does not mean they were not critical of them at the same time. In fact, they often tried to exert pressure on administrations so that they effectively safeguarded and promoted private initiative, which the gabinete owners regarded as crucial for the growth of their industry. In generally cooperating with public institutions while at the same time trying to keep them in check, the *gabinetes* were also in line with strands of *regeneracionista* thought. In February 1900, Ureña and other unnamed *gabinete* owners sent a letter to the ministry of Finance to inquire as to why import taxes for phonographs and cylinders had risen by 1,000% following the recent tax reform.[43] There is no evidence that the ministry of Finance ever responded to Ureña, and the same can be said about further complaints put forward by the industry in the following years about the Spanish taxation system and the disadvantages it posed to *gabinetes*.[44] It is likely, however, that by publicizing their complaints about taxation the *gabinetes* were not just targeting the tax office itself: they were also presenting themselves as socially aware, patriotic entrepreneurs to readers—and therefore potential customers.

The fact that many *gabinetes* were keen to cultivate and promote connections with the provinces is a further sign of their *regeneracionista* commitment to their country, since the lack of transport connections and the rivalries between regions were frequently cited, among *regeneracionistas*, as one of the causes of Spain's backwardness. Many *gabinetes* proudly claimed in their advertisements that they sent their products to the provinces on demand (e.g., Viuda de Aramburo, La fonográfica madrileña, El graphos), and they also advertised in newspapers outside Madrid,[45] whereas others had permanent representatives in one or more Spanish provincial cities (e.g., Fono-Reyna in Logroño).

Recording and labor in *regeneracionista* Spain

Thus far I have discussed how the nascent identity and discourse developed by owners of *gabinetes fonográficos* fitted in within existing narratives of science, business, and patriotism connected to *regeneracionismo*. It could be argued, nevertheless, that the previously discussed strategies were not specific to the recording industry, but were indeed shared with other entrepreneurs no matter what industry they worked in. In this section, I turn my attention instead to industry-specific issues, and particularly to how *gabinetes* developed the recording as a commodity following global developments, but also adapting these to their cultural context and local needs. During the era of traveling phonographs, the device's appeal was in its ability to reproduce reality, and not so much in providing the aesthetic enjoyment that could be procured by a professional singer singing well-known numbers of *zarzuela* and opera in one's own home: we might say that the recording, as we know it, did not exist yet at this time, and it was the task of the *gabinetes* to develop it as such.

Of course, it cannot be established with total accuracy when this notion came into being, or when it was accepted by a sizeable part of the Spanish population: in fact, as will be discussed in Chapter 4, the development of the recording as a community in Barcelona lagged behind Madrid, for reasons having to do with the particularities and penetration of the industry in each of these cities. Advertisements, as well as writings in newspapers and in *El cardo* and recordings themselves, allow us to examine how *gabinetes* spoke of their products and technologies, and how the way they did so changed over time. These changes suggest that the notion of the recording as commodity became commonplace in the *gabinetes*' discourse in a relatively short period of time between approximately 1897 and 1899–1900. When Hugens y Acosta started advertising in the press in March 1897, their advertisements did not make it clear whether the *gabinete* was selling its own recordings. However, there is a strong likelihood that that this was not the case: with Hugens having a background in organizing phonograph demonstrations and sessions, we might presume that he initially stuck to that in his newly found *gabinete*; we might also hypothesize that, had he produced his own material, he would be keen to promote it. When Viuda de Aramburo started to advertise in newspapers in May 1897,[46] they already mentioned "impressed Spanish and American cylinders" in their advertisements. It is likely that Hugens was recording his own cylinders at this stage or started to do so shortly thereafter, as he tended to always keep up to date on the developments of the industry. Nevertheless, at this point and throughout 1897, the publicity of the few existing *gabinetes* still focused on phonographs, graphophones, and accessories rather than on recordings. Recordings were indeed mentioned, but advertisements did not include any details at this stage as to what sort of repertoire they consisted of, who recorded them, or how much they cost. It is plausible that the ten Viuda de Aramburo cylinders available today with no performer attribution (out of a total of 105) date from these early days, with the *gabinete* then making it the norm to include the names of the performers. We might infer from this that the appeal, therefore, was still in the machines themselves and their ability to reproduce sound (as was the case in the era of the traveling phonographs), and not so much in collecting and consuming individual recordings of specific pieces, or by specific performers.

Within two years, the situation had changed. The first known catalogues, dating from 1899 and 1900, were structured around recordings and singers and prominently displayed their names and photographs, but not always the titles of the pieces they recorded. Some *gabinetes* still released some of their cylinders without indication of the performer as late as 1900–1905, but evidence indicates that this was more common in the case of instrumental music and opera and *zarzuela* choirs, who typically were less popular with audiences than opera or *zarzuela* arias and ensembles: soloists were, as a norm, properly identified in the box and/or recording itself.[47] Together with the catalogues, this suggests that the notion of the recording as a commodity had started to enter the Spanish consciousness: at least some Spaniards were now starting to become interested in buying recorded music as an aesthetic artifact, involving recordings of specific pieces of music performed by specific musicians, and not purely to reassure themselves of the ability of the phonograph to reproduce reality, as had been the case throughout the 1890s.

It is not only catalogues and advertisements that help us understand how the notion of the recording as a commodity evolved in Spain: the very recordings themselves also offer very important clues. While phonograph demonstrators throughout the 1890s only needed to make sure that their recordings were a satisfactory enough reproduction of reality, *gabinete* recordists were now faced with having to make decision about how to organize and present recorded sound, what quality standards to develop and aspire to, which features and characteristics of the music to include in the cylinders and which ones to leave out—in other words, they needed to codify the phonograph recording as a genre. Decisions Spanish *gabinetes* needed to take included, for example, whether to include solely music or other kinds of sounds too (they typically did the latter, under the form of a spoken announcement at the beginning and applause at the end), or whether to split longer pieces into several recordings, therefore breaking the illusion of a continuous performance (which they overwhelmingly did not). Commercial recordists elsewhere in the world were doing the same at this time, engaging, in the words of Patrick Feaster, in "a creative and emergent process, not a straightforward borrowing of established musical and narrative categories" as they codified the recording as a genre.[48] In the remainder of this section, I discuss how *gabinetes* uniquely shaped this process and how they made themselves and their labor visible and audible both in the discourse surrounding the recordings and in the recording themselves, while keeping in line with their *regeneracionista* aims.

The most literal way in which the *gabinete* owners made themselves audible in recordings concerned no doubt the practice of saying out loud, at the beginning of the recording, the name of the performers (but not normally the accompanist, in the case of singers), the piece they were playing or singing, and (finally, perhaps more importantly) the name of the *gabinete* they were recording for. The announcement, which appears in practically all the commercial recordings of the time made in Spain,[49] would normally be made by the operator making the recording. In most cases, the announcer's voice reappeared at the end of the recording, clapping and bravoing the performers enthusiastically. The practice of announcing the names of the musicians, title of the piece, and company name was commonplace in other countries too; Feaster argues that the announcement was not simply intended as one additional way to identify the contents of the cylinder (e.g., if the cylinder did not come with label, or out of concerns that the label would deteriorate over time), but was instead an essential part of the phonogenic enactment.[50] In this context, announcements instead provided proof that the phonograph could handle the human voice before the actual song was played, therefore reassuring listeners; it could also be intended as an autograph, signifying that the recording was not the product of the labor of the performers alone, but of the recordist too.[51]

Both of Feaster's hypothesis can be productively applied to the early Spanish cylinders. *Gabinetes* were keen to show that the phonograph could indeed handle the human voice and reproduce it to a level of fidelity that was acceptable for the audience (a remnant of the earlier era of traveling phonographs), and at the same time they were keen to signal that this was not something that the phonograph could achieve on its own, but it required hard work and know-how from the operator[52]—for example, in terms of positioning the singers and pianist in certain ways as to achieve the best balance, or in terms of choosing the right diaphragm among several available options (some manufactured abroad and imported, some developed by the *gabinete* itself). For consumers familiar with the *gabinetes'* publicity, their publications, and their discourses, it is likely that these announcements functioned as a reminder of the particular set of know-how, skills, and tacit knowledge that they thought *gabinete* operators to have, often developed through trial and error and possibly too through exchanges with colleagues and with customers themselves who made their own home recordings.[53] Many customers would be aware of the efforts of *gabinetes* to introduce innovations and additions to the recording process and recording devices, as the *gabinetes* themselves normally went to great lengths to publicize such efforts both in the press and on their premises. The development of such innovations was therefore not simply necessary for commercial purposes: the *gabinetes* also regarded it as a key part of their developing identity under the *regeneracionista* banner. The *gabinetes* were indeed keen to show that they were not simply concerned with disseminating and distributing existing inventions and technologies among the Spanish population, but were also well practiced in developing and improving technology themselves. Most such developments and improvements pointed in the direction that the future of the phonograph was in entertainment rather than in business and bureaucracy, as had been Edison's original intention; for example, Fono-Reyna was credited with inventing a special microphone for making recordings,[54] and La fonográfica madrileña also patented a new recording process.[55] Innovations to develop the phonograph's business uses, on the other hand, were almost non-existent at the time in Spain.

Another way in which the *gabinetes* made themselves and their labor visible and audible was through engaging in and even actively starting debates and controversies regarding their difficulties in navigating the restrictions imposed by Spanish law or the Spanish economy. This would, again, confirm their role as *regeneracionista* entrepreneurs developing an innovative product who nevertheless sometimes clashed with the passivity of the institutions. These controversies mostly appeared in *El cardo*, but they occasionally reached the general press through letters to the editor. In doing so, the *gabinete* owners were attempting to position themselves within nationwide concerns, presenting problems specific to the nascent industry as issues of interest to a broader section of the Spanish population—that is, those with enough cultural capital to read newspapers and other publications and keep up to date with economic, social, and cultural developments in the country.

The clearest example of this concerns the issue of cylinder duplication, which remained a controversial topic throughout the *gabinetes* era in Madrid. As I have explained earlier, Spanish cylinders could not be easily duplicated at the time, but some *gabinetes* tried to do so by using a machine akin to the pantograph or by having singers record for more than one phonograph at the same time; others tried to develop new technologies and techniques to make duplication possible. While some of the most visible *gabinete* owners and other significant voices in the industry opposed cylinder duplication on the grounds that it dramatically affected sound quality, the fact that others were actively trying to make it possible suggests that the industry was divided on this issue or at the very least that opinions around cylinder duplication were ambiguous: some *gabinete* owners might have regarded it as a potentially appealing way of increasing their output and hence their sales, but they might have felt at the same time that the duplication methods available to them were not up to standard. Tellingly, in an article in *El cardo* signed with his real name, the Marquis of Alta-Villa referred to the cylinder duplication issue as "la cuestión palpitante" (the burning question)—a phrase which had been famously used by Spanish novelist Emilia Pardo Bazán to title an essay collection in 1882 in which she reflected about realism and naturalism. Alta-Villa was therefore suggesting that cylinder duplication was as serious a cultural issue as the representation in literature of social problems and dynamics had been twenty years earlier[56]—and, as such, concerning not only those in the scene (writers and readers in the case of literature, *gabinetes* and consumers in the case of recording technologies) but also Spanish society more broadly.

The first significant example of how some *gabinetes* used the cylinder duplication issue to attract attention to themselves and advance *regeneracionista* discourses dates from early 1900. At this point, a group of singers allegedly met with officers at the ministry of Economy to plead with him that the duplication of recordings be made illegal. Our only source for these meetings, though, is an article that Álvaro Ureña wrote in *La correspondencia militar*,[57] and not records written by singers themselves or by civil servants; this means that we should be wary of how Ureña represented the issue. His article indeed leaves some unanswered questions. First of all, it does not fully explain how some *gabinetes* achieved cylinder duplication at this stage. What he says is that the singers had complained that the cylinders they recorded were duplicated at a later stage *without them knowing*. This suggests that duplication was achieved here by mechanical means, and not by the singers being made to sing concurrently for more than one phonograph. Secondly, it is not entirely clear what the singers' demands were, but Ureña's reply suggests that the singers' objections to this practice were financial rather than artistic. *Gabinetes* normally paid singers a pre-arranged fee for each cylinder they recorded, and so unauthorized duplication meant that *gabinetes* were profiting from the singers' labor without appropriately compensating them.

In his reply, Ureña chose to speak not only for himself but for the rest of the industry, claiming that only Viuda de Aramburo and Hugens y Acosta sold duplicated cylinders, but himself and most other *gabinete* owners would never "dream of doing so."[58] In doing so, Ureña became one of the first to openly voice the notion that the *gabinetes'* industry was based on artisanal labor instead of mass production; it therefore retained a personalized, exclusive touch that tied in well with the notions of luxury I have discussed earlier.

Ureña's article did not immediately receive a reply either from singers, civil servants, or other *gabinetes*, and there is no evidence that any action was taken by the Ministry of Economy against the duplication of cylinders. As with other letters and articles that *gabinete* owners sent to the general press, this can be regarded as a strategy Ureña followed to get what he perceived to be the industry's concerns and issues out in the public sphere, rather than a measure intended to result in specific action. This was successful to some extent, as the issue of cylinder duplication kept coming up in the general and specialized press—particularly in *El cardo*. This is understandable because in the years 1900–1901, threats to artisan cylinders were coming not only from cylinder-duplicating machines but also from the increasing popularity of the gramophone, whose discs could be duplicated with considerably more ease. *El cardo*'s and Ureña's activism against duplication must therefore be read as a further attempt from the industry to make their labor clear and audible to audiences, suggesting that each recording was unique and had been carefully looked after by a phonograph operator, and not simply generated through mass production processes. This was, again, connected to broader *regeneracionista* debates. *El cardo* repeatedly claimed that artisanal production of cylinders should be a matter of national pride, and argued that Spanish *gabinetes* were much more gifted at crafting cylinders than their foreign counterparts were—but the vigor of the industry could only be maintained if the *gabinetes* avoided cylinder duplication and instead made sure that each cylinder was manually crafted.[59] Cilindrique stated that Spanish cylinders had admirable color, sonority, and clarity, and the Frenchmen, German, and Italians could indeed learn the technique from the Spaniards.[60] Comments about the alleged superiority of Spanish cylinders over those produced elsewhere in Europe became commonplace in *El cardo*,[61] sometimes veering on the chauvinistic: an anonymous article in January 1901 stated that, whereas Spain exported only 1,000 cylinders per month (compared to 60,000 from Italy and 2 million from France), French cylinders were all "dire" whereas all the Spanish ones were very well impressed.[62] The magazine also regularly printed stories—real or apocryphal—of foreign audiences becoming hugely impressed by the technical standard of Spanish cylinders—for example, with Armando Hugens traveling to France and playing back his own cylinders there to the astonishment of audiences.[63] Ureña himself claimed in 1901 that the Spanish phonographic industry was on a par with that of the rest of Western Europe—a highly significant claim to make at a time in which Spain as a whole was struggling to keep up with the rest of Europe in other respects—but, contrary to what happened in other countries, the Spanish government was not granting any sort of protection to the industry.[64]

A direct reply to Ureña's article took more than one year to materialize—in the previously mentioned article "La cuestión palpitante" by the marquis of Alta-Villa. Here, Alta-Villa claimed that Viuda de Aramburo and Hugens y Acosta were indeed guilty of duplicating cylinders. He did not provide any evidence emerging from the actual recordings, but claimed that, if Ureña's accusation was untrue, Viuda de Aramburo and Hugens would have defended themselves, which they did not: instead, they did not reply to Ureña's letter in any way.[65] In response to the singers' demands, he suggested that they singers should start a trade union to protect themselves, but he claimed that the main problems with duplication were not financial, but rather concerned artistic quality. He suggested that sellers should examine very carefully whatever they bought to make sure that it was not a copy.[66] As pointed out by Alta-Villa, Hugens y Acosta and Viuda de Aramburo did not reply to either Ureña's original article or to Alta-Villa's follow-up. This is interesting in itself, as it suggests that the anti-duplication position carried with itself more prestige and acceptance within the community than the pro-duplication one: Hugens and Viuda de Aramburo might not have felt they could justify their business strategies while staying within the *regeneracionista* ethos in the Madrid industry. Things were different elsewhere, though: even though *El cardo* and Ureña often claimed to be speaking for the Spanish recording industry on the issue of cylinder duplication, the reality was that not all *gabinetes* all over the country attached the same value to the artisanal production of cylinders. Specifically, Valencia *gabinetes* and their magazine, *Boletín fonográfico* (discussed in Chapter 5) dedicated considerable efforts to the development of techniques for cylinder duplication, and they did not voice the same kinds of arguments for artisanal duplication as their Madrid counterparts did. Indeed, whereas articles praising the artisanal production of cylinders were commonplace in *El cardo*, they hardly feature in *Boletín fonográfico*. *Boletín fonográfico* even gave significant coverage to *gabinete* Pallás when its owner announced that he had found out how to reproduce cylinders, which would allow them to sell cylinders at just 1.5 *pesetas*.[67] *El cardo*, on the other hand, dismissed Valencian cylinders and claimed that they suffered from a lack of craftsmanship,[68] although at a later stage the magazine conceded that Valencian *gabinetes* had managed to catch up with those in Madrid in terms of recording quality.[69]

Another, shorter-lived controversy the Madrid *gabinetes* engaged in concerned copyright. Although less significant than the discourse around cylinder duplication, this too speaks of how *gabinetes* conceived of their own efforts in shaping the recording as a commodity. In November 1900, *El cardo* criticized an employee of Viuda de Aramburo for having "rushed," almost immediately after the *gabinete* opened, to obtain from *zarzuela* authors the permission to record their works exclusively. This would in practice mean that numbers from certain *zarzuelas* (presumably the most popular ones, even though *El cardo* did not give details) could only be recorded by Viuda de Aramburo. *El cardo* implied that this was only possible because the employee had taken advantage of a "grey area" in Spanish law concerning copyright. In the same article, Ureña was reported to have signed similar agreements with composers, but was still happy to give his permission to record these authors' works "to everybody who records these same authors under the same conditions as Ureña does."[70] It is not clear what Ureña meant by "under the same conditions," but, given his interest in the non-duplication of cylinders, it is likely that what he had in mind is that other *gabinetes* should not duplicate the cylinders they recorded of the composers concerned. It might also be that he expected other *gabinetes* to pay singers a similar wage as he did and thus to sell the resulting cylinders at a comparable price, to avoid undercutting.

The surviving cylinders and catalogues do not provide strong evidence that the alleged exclusive contracts signed by Ureña or Viuda de Aramburo were enforced in practice. Neither *gabinete* claimed in its wax cylinder cases that recordings were exclusive, and, since no catalogues of either have survived, we cannot know if they made that claim in those. The repertoire included in surviving cylinders paint a similarly inconclusive picture: we might presume that Ureña or Viuda de Aramburo would have signed contracts with the most successful *zarzuela* composers, as these were the most likely to bring in a profit. These would have surely included Manuel Fernández Caballero and Ruperto Chapí, who also happen to be the *zarzuela* composers of whom the most cylinders have survived (30 for the former and 34 for the latter).[71] However, the surviving cylinders of these two composers come from a range of *gabinetes* all over the country. The same can be said of other similarly popular composers of whom fewer cylinders have survived (Federico Chueca, Tomás Bretón); in their cases, the range of *gabinetes* is comparatively narrower, but it does not suggest that either of these might have had exclusive contracts with any of the previously mentioned *gabinetes*. Moreover, with some *gabinetes* not issuing catalogues and selling a range of cylinders which were for the most part unique and did not have at the time a matrix or registration number, to effectively ensure that certain composers were not being recorded by certain *gabinetes* would not have been feasible unless constant monitoring of recording sessions and cylinders was put into practice. Nevertheless, even if the copyright status of phonograph recordings was still unclear at this time and no measures were taken to regulate the issue, the fact that some *gabinete* owners were engaging in debate concerning copyright and exclusive contracts again suggests that they regarded this as another area in which to leave their footprint on recorders. *El cardo*'s and Ureña's insistence that every *gabinete* should be free to record any composer they wanted suggests that they trusted that the main differences between two recordings of two different *gabinetes* should not be in the music itself, but rather in the expertise and know-how demonstrated by the operator in recording the music. *El cardo*, nevertheless, eventually conceded that exclusive contracts could provide *gabinetes* with an opportunity to differentiate their products from others', with exclusive contracts offering enhanced guarantees to customers that cylinders were authentic.[72]

Conclusion: End of an era

The last *gabinete* to still sell its own cylinders, Hugens y Acosta, liquidated its assets in 1905; as Hugens y Acosta was too the first *gabinete* to open in Madrid and in all of Spain, this can be rightly regarded as the end of the *gabinete* era. Up until that point, the penetration of multinational companies into the Spanish market and the replacement of phonographs with gramophones had happened gradually, and was initially saluted with skepticism by some. In February 1901, *El cardo* wrote that: "The delicacy of the phonograph captivates, charms and seduces; the energy and the big sound of the gramophone make it a precious instrument, which satisfies and reproduces, very close to the truth, certain kinds of sounds and effects."[73] The rhetoric suggests, as is typical of this publication, suspicion against industrial production of recordings, but it also conveys a certain determinism, conceding the gramophone an advantage over the phonograph, but at the same time lamenting the potential loss of some qualities of the phonograph. Later that year, in September, *El cardo* referred to an "engineer" sent by a "London company" (referring to Gramophone) to Madrid and producing substandard material with "mediocre" singers.[74]

When Gramophone opened an office in Barcelona in 1903, it was not long before some Madrid *gabinetes* became its agents: La fonográfica madrileña was selling gramophones and discs by April 1903,[75] and Ureña was doing so by the end of the year.[76] The shift from artisanal to industrial was relatively quick, but gradual, and even in 1903 there was some capital attached to the artisanal production of cylinders: Fono-Reyna claimed in its advertisements of that same year that it was "the only business in Spain which is consecrated to the *artistic* impression of direct cylinders and does not duplicate them."[77] Even though Gramophone remained the most powerful multinational company in Spain during these years, others managed to introduce their products too, such as French-based Odeon and Italian Fonotipia, whose records were being sold by La fonográfica madrileña in 1905;[78] unlike in other contexts like Scandinavia,[79] though, local companies producing their own records could not thrive in Spain next to multinationals— perhaps because Spain, together with Latin American territories, was sufficiently large as a market that it was profitable for multinationals to fill all available niches in terms of repertoires. In any case, the Madrid *gabinetes* left a significant legacy that surely paved the way for multinational companies: they introduced in Spain to a greater extent than *gabinetes* elsewhere, the notion of the recording as a commodity that could be bought and sold and positioned this notion within debates of *regeneracionismo*, science and technology, and national identity—debates which in turn shaped aurally some of the conventions of the nascent medium.

4

Science, urban space, and early phonography in Barcelona, 1898–1914

The development of domestic phonography and record consumption in the early years of the twentieth century caught Barcelona in the middle of a decades-long transformation from provincial capital into Mediterranean metropolis. The transformation affected the very morphology and size of the city. The medieval walls were gradually demolished starting from 1854; with this, Barcelona began a process of expansion, urbanization, and rationalization not dissimilar to the equivalent processes other major European cities were undergoing at the time. Population grew too, and in 1900 Barcelona could boast 537,354 inhabitants to Madrid's 540,109. Many of these new inhabitants were manual workers coming to Barcelona from elsewhere in Catalonia and Spain to work in the developing textile industry. Inequality was thus on the rise and would, in the first decades of the twentieth century, degenerate into violent social confrontation—the most famous example being the 1909 Semana Trágica (Tragic Week). These profound changes not only gave rise to Barcelona's international reputation as a modern, progressive city (a reputation that has survived to our days) but they also redefined its significance as the capital of an increasingly industrialized region whose rising, thriving nationalist movement often positioned itself in opposition to Madrid, arguing that the former was more modern, industrialized, and European than the latter.

We might presume that Barcelona's preoccupation with science and modernization as a core part of the city's nascent identity, as well as its ongoing urbanization, would have resulted in rapid and widespread adoption and development of recording technologies after domestic versions of the phonograph were introduced from 1896 onward. Nevertheless, even a cursory look into the city's early recording industry reveals much less vibrancy than in the capital of Spain. Against Madrid's fourteen *gabinetes*, Barcelona had only nine, and most of those were only active for a very short period of time—which in some cases amounted to mere months. Even though I have already discussed in Chapter 3 how the number of surviving cylinders does not necessarily provide an accurate picture of how many recordings would have actually been produced, the Madrid cylinders held between the Biblioteca Nacional de España, Biblioteca de Catalunya, Museu de la Mùsica, and Centro de Documentación Musical de

Inventing the Recording. Eva Moreda Rodríguez, Oxford University Press. © Oxford University Press 2021.
DOI: 10.1093/oso/9780197552063.003.0005

Andalucía clearly outnumber those from Barcelona: some 180 come from the former, and only 53 from the latter.

Besides being less significant numerically and commercially, the Barcelona *gabinetes* also failed to articulate a discourse concerning their own identity and place in national debates and concerns in the same way as did those from Madrid or Valencia (the latter of which will be discussed in Chapter 5). Contrary to Madrid's *El cardo* and Valencia's *Boletín fonográfico*, the Barcelona *gabinetes* did not even have a dedicated publication they could use to showcase their products and innovations, defend their interests, and develop a discourse aimed at placing them and their products in the center of national (be it Spanish or Catalan) debates and concerns. At the same time, neither *El cardo* nor *Boletín fonográfico* occupied themselves very extensively with record production in Barcelona, which also suggests that Catalan *gabinetes* were not perceived to be very significant even by their colleagues elsewhere. Writers in *El cardo* claimed to be aware of *gabinetes* in Barcelona, but only superficially,[1] and they only ever named one Barcelona establishment individually: Corrons, who, in the opinion of the writer, manufactured blank cylinders whose quality was on a par with those imported from abroad.[2] Valencia's *Boletín fonográfico* referred on a few occasions to the Barcelona *gabinetes* as a group, although, again, only Corrons was named individually.[3] As will be discussed later, Corrons was likely the longest-lived *gabinete* in Barcelona, and half of the surviving Barcelona cylinders came from this establishment, which might indicate that it was the most productive too—and hence the most likely, among the Barcelona *gabinetes*, to have acquired a customer based outside its indigenous city and thus have gained the attention of *El cardo* and *Boletín fonográfico*.

In the same way as the Barcelona *gabinetes* did not have their own publications, there are also no known instances of them sending letters to the editors of newspapers or otherwise attempting to turn the concerns of the industry into a matter of national importance (either for Spain or for Catalonia). No catalogues from the Barcelona *gabinetes* have survived either, which, in the case of the Madrid and Valencia establishments, were an important part of establishing the *gabinetes'* identity as *regeneracionista* entrepreneurs and developers and manufacturers of a range of accessories and improvements for the phonograph, as opposed to mere resellers of Edison's products.

This chapter begins with an overview of the available evidence about *gabinetes fonográficos* in Barcelona. I then discuss this evidence to advance two non-mutually-exclusive hypotheses as to why these *gabinetes* did not reach the production levels or cultural significance of some of the *gabinetes* in Madrid. Both explanations have to do with the failure of the *gabinetes* to integrate themselves, on the one hand, in the changing geography of the city and, on the other, in developing discourses about science and modernization. Lastly, I examine how the situation shifted after the arrival of multinational companies between 1903 and 1906, with recording culture in Barcelona thriving in close connection to the main centers of record production in Europe, while at the same time developing local particularities which strengthened the city's position in these international networks. Building on Chapter 3, this chapter intends to further exemplify how place, through both its very materiality and successive discourses attached to it, shaped how recording technologies and recordings were used and woven into everyday life and dominant discourses, and how place was in turn shaped by them.

The Barcelona *gabinetes*

Compared to Madrid, phonographs for domestic use and wax cylinders made a somewhat belated appearance in Barcelona. The first advertisement in the local press dates from February 1899, when the department store Grandes Almacenes El Siglo announced that its opticians' section was now selling phonographs and accessories, as well as cylinders. No details were given as to whether the cylinders' were produced by El Siglo itself or imported, what type of repertoire they contained, or which singers or instrumentalists were featured in them.[4] The first notice we have of a *gabinete* proper—that is, a small business dedicated to re-cording their own material and selling imported phonographs—dates from April 15, 1899.[5] This was date of the launch of the Sociedad Artístico-Fonográfica, which organized an event for the occasion that is certainly reminiscent of the launches of some of the best-known *gabinetes* in Madrid. The owners of Sociedad Artístico-Fonográfica, Juan B. Baró and Juan A. Rosillo, enlisted the help of well-known Barcelona opera singers soprano Josefina Huguet (a regular of *gabinetes* all over Spain who then went on to record for Gramophone and Victor), tenor Ángel Constantí, and baritone Mr. Puig. As was the norm in the demonstrations of touring phonographs during the early and mid-1890s and the inauguration events of other *gabinetes*, the arias the singers performed were recorded and then played back immediately afterward. With this format being commonly used to draw attention to the phonograph's capabilities to reproduce audible reality, it is not surprising that the anonymous reporter for *Los deportes*—the only extended review of the event—wrote that: "We were highly impressed by the clarity and purity of every note, even the softest ones, as they were played by the Sociedad's perfected phonographs. There was no trace of the metallic shrieks of the vulgar phonographs that we are all familiar with."

This last sentence speaks of a wish from the part of the Sociedad to distance themselves from the *salones* that I have discussed in Chapter 2 and that would have presumably been less reputable than *gabinetes* such as Hugens y Acosta,

Álvaro Ureña, and Viuda de Aramburo. Among the owners of Barcelona *gabinetes*, Baró and Rosillo were indeed, from the beginning, the most active at attempting to build the same sort of reputation that some of the Madrid *gabinetes* had already acquired. Within two weeks of the official inauguration, the Sociedad held phonographic sessions at the Teatro El Dorado[6] and the Teatro de Cataluña,[7] presumably in an attempt at publicizing itself and its products. It is likely that such sessions followed the same format as the earlier phonographic demonstrations, with music and the spoken word being recorded and then played back so as to demonstrate the capabilities of the devices. The phonograph used in the Barcelona performances of Chapí's *El fonógrafo ambulante* in May 1899 was also on loan from the Sociedad.[8]

Baró and Rosillo were also keen to position themselves not simply as mere importers of Edison's inventions or as businessmen primarily dedicated to impressing cylinders, but also as inventors in their own right. In the weeks following the inauguration of their *gabinete*, the general interest magazine *La hormiga de oro* reported that the Sociedad had developed an "artificial membrane" (likely a diaphragm) that would make the recording process easier.[9] The industry and science publication *Industria e invenciones* reported that Baró and Rosillo had invented a new microphone, as well as a method that would allow the company to record a full act of an opera.[10] While the former claim is perfectly plausible (Corrons also patented a highly successful microphone, as I will discuss later), the latter claim is more dubious. No such full recordings of opera acts have survived, and, with the recording industry struggling at the time to make longer recordings, Baró's and Rosillo's invention, if commercially viable, would have been a global breakthrough that other countries would have no doubt rushed to adopt. As with the Madrid *gabinetes* and inventors, the news stories about new technological developments carried out in Barcelona must not necessarily be taken at face value, but rather as evidence of how *gabinetes* tried to establish their technological credentials in an attempt to acquire status and costumers.

Although comparatively little is known about Rosillo, evidence suggests that he was the closest that Barcelona might have had to Armando Hugens, Álvaro Ureña, Atanasio Palacio Valdés, or the Marquis of Alta-Villa—men who believed in the long-term potential of the phonograph and had scientific, technical, and commercial ambitions, as well as some artistic sensitivity. Rosillo filed five patents related to the phonograph; the first one (no. 23185), for a microphone, in October 1898; he then introduced some improvements to the patent in February 1899 (no. 23781), and filed a second one in April 1899 (no. 25064) for modifications to the Edison phonograph. The three remaining patents were filed after he closed his *gabinete*,[11] which is surprising. Indeed, for all its initial impetus, the Sociedad Artístico Fonográfica appears to have been short-lived: no advertisements or references can be found in the press after June 1899, and only one cylinder of the Sociedad has survived.[12] Evidence suggests that Rosillo was involved in another *gabinete*, Depósito General Fonógrafico y Fotográfico, later in the same year.[13] The establishment had been active as a photograph shop under the name Depósito General Fotográfico until approximately July 1899.[14] With no further evidence available, we might hypothesize that Rosillo initially planned to run the Depósito General Fotográfico and the Sociedad Artística Fonográfica as separate ventures, but subsequently, at some point between July and October 1899, decided to bring them together on the premises of the former as Depósito General Fonográfico y Fotográfico. Even though no cylinders from Depósito have survived, there is evidence that the *gabinetes* remained active for a few years: in 1901, the Depósito had a representative in Alcoy, Valencia,[15] and by 1903 it was in the hands of merchant Ramón Olaguer-Feliu according to *Anuario del Comercio*. With the gramophone quickly imposing itself in Barcelona and the rest of Spain around that time, it is plausible that Depósito closed shortly after that.

It is likely that it was the launch of the Sociedad Artística Fonográfica that encouraged other shop owners and entrepreneurs in Barcelona to open their own *gabinetes*, with several new openings clustering in the months following April 1899. In May, Roselló, an optician, was advertising Spring motor, Home Standard, Columbia Victoria-Eagle phonographs, as well as Bettini diaphragms and blank and impressed cylinders.[16] Like many other *gabinete* owners, Roselló had a track record of introducing other technological and scientific innovations in Spain, such as lightning rods.[17] In October 1899, the well-established musical store Ribas y Estradé, which sold pianos and other instruments, started to advertise phonographs and cylinders too.[18] Juan Bautista Estradé Simó and Ricardo Ribas Anguera, the owners, had been seemingly interested in phonography for at least several months before starting operation, and in February 1899 they teamed up with Jaume Ferrán Clúa, a medical doctor and bacteriologist and amateur photographer, to develop a patent (registry no. 23700) introducing innovations to the manufacturing of wax cylinders—a further example of the *gabinetes'* concern to establish their credentials as inventors and developers in their own right.[19] Neither the Sociedad nor Roselló or Ribas y Estradé included any details in their advertisements of the sort of repertoire that was on offer, or the singers who recorded for them, which suggests that—as was the case with the first *gabinete* advertisements in Madrid—the appeal for customers at this stage was in the scientific capabilities of the phonograph, rather than in the possibility of listening to specific singers or pieces.

Corrons—the most successful of the Barcelona *gabinetes*—likely started operation at some point in 1899 as well. The Corrons family was a well-known lineage of opticians, and some of them had introduced various scientific innovations in Spain since at least the 1860s.[20] The first cylinders of the Corrons *gabinete* were released under the name "V. Corrons," or Viuda de Corrons—referring to the widow of optician Joaquín Corrons, who ran her late husband's business between his death in late 1898 and her own on January 20, 1900. After this, the business was taken over by their son, José, and wax cylinders produced thereafter were labeled "J. Corrons."[21] José Corrons moved his father's business to new premises and started advertising himself as a seller of scientific equipment, as well as phonographs and cylinders.[22] In a memoir, Corrons himself wrote that his effort to popularize phonographs and cylinders in Barcelona became so strenuous that he became "sick with neurasthenia."[23] Evidence certainly suggests that Corrons was one of the most active *gabinete* owners, not only in Catalonia but also in the rest of Spain: his was one of the establishments of which the most cylinders have survived, and he was also the only *gabinete* in Barcelona to consistently advertise and be mentioned in *Boletín Fonográfico*, indicating that his reputation certainly spread outside his native city. Corrons also achieved a modicum of recognition among record collectors in Barcelona and beyond for a microphone of his own invention, the *micrófono Corrons*, which was duly mentioned in the announcements at the beginning of his recordings ("impresionado con el micrófono Corrons").[24] In mid-1901, together with a partner named Canals, Corrons also started manufacturing blank wax cylinders.[25] At some point, Corrons must have started selling Gramophone discs too: a Gramophone catalogue of Portuguese, French, Italian, and Spanish records dating from around 1900 has been preserved at the Biblioteca Nacional de Catalunya with Corrons's stamp on the cover.[26] The *gabinete* remained active until around 1903: in the *Anuario del Comercio* 1904, Corrons appears as being operative in the areas of photography and optometry only, but not phonography.

The last *gabinetes* to open in Barcelona include Manuel Moreno Cases, who opened his Centro Fonográfico Comercial in December 1900,[27] advertising Concert phonographs for 500 *pesetas*, gramophones for 110, and impressed cylinders; and O. Pernat, who opened his in January 1901.[28] Finally, nine cylinders have survived from a *gabinete* named El fonógrafo, owned by optician Conrado Olió.[29] Since he did not publish press advertisements or date his cylinders, it is not possible to ascertain the establishment's dates of operation. Therefore, among all the Barcelona *gabinetes*, there is only evidence that Depósito General Fotográfico y Fonográfico and Corrons were operative for more than one year. Their owners were also the most active in promoting their establishment as well as their scientific and/or commercial credentials; however, in comparison to Hugens, Ureña, and the Valencia *gabinetes*, their efforts, as well as their awareness of the social and cultural context they operated in, remained limited.

In the absence of direct evidence from the *gabinete* owner themselves (with the exception of Corrons) or from industry publications originating in the city,

Table 4.1 *Gabinetes fonográficos* in Barcelona

Name	Dates active[a]	Key individuals	Own recordings?	Other activities	Other
Centro Fonográfico Comercial de Manuel Moreno Cases	1900	Manuel Moreno Cases	Yes	Cinema	
Centro Fonográfico O. Pernat	1901	O. Pernat (?)	Yes	Clockmaking	
Depósito General Fotográfico y Fonográfico	1899–1904 (?)	Juan A. Rosillo, Ramón Olaguer-Feliu		Photography	
El fonógrafo	1898 (?)–1904 (?)	Conrado Olió	Yes	Optician	The dates 1898–1904 are given by BDH, but they do not come from either the cylinders themselves or press advertisements
Grandes Almacenes de El Siglo	1899	Eduardo Gómez, Ricardo Conde, Pablo del Puerto	Yes	Department store	
Ribas y Estradé	1899	Juan Bautista Estradé Simó, Ricardo Ribas Anguera	?	Music shop	
Roselló	1899	Roselló, Widow Roselló	Yes	Optician	
Sociedad Artística Fonográfica	1899	Juan B. Baró, Juan A. Rosillo	Yes	None	
V. Corrons, J. Corrons	1899–1904	Widow Corrons, Joaquín Corrons	Yes	Optician	V. Corrons in 1899, J. Corrons from 1900 onward

[a] "Dates active" refers to the dates for which we know (through press advertisements and articles, or through inclusion in industry yearbooks) that a particular establishment was operating as a *gabinete fonográfico*, that is, selling phonographs and/or recordings. The establishment may have been active selling other types of products or services before or after that.

it is difficult to provide a conclusive answer as to why the Barcelona *gabinetes* never reached the success of those in Madrid and Valencia. One reason might have to do with the personalities and motivation of the individuals involved. Whereas the Madrid industry was driven by a few very motivated, active individuals, such as Hugens, Ureña, and Alta-Vila, none of the *gabinete* owners in Barcelona—perhaps with the previously mentioned exceptions of Rosillo and Corrons—seemed to have the same stamina and resilience, or the same talent to lead and inspire others. There are other potential reasons, however, that go beyond individual personalities and interweave the back and forth of the nascent industry and the cultural and social conditions it found in Barcelona. Indeed, collectively, the Barcelona *gabinetes* also lacked the ability to play to their advantage two sets of discourses that dominated life in the city at the time: one on urban change and spatiality, and the other on science and progress. In the historical and cultural context at hand, these two sets of discourses were inextricably connected to one another under broader issues of modernization and national identity. Advances in science and technology made it possible for cities to grow and be organized and be managed according to rational principles. Technology and applied science also allowed certain social classes to amass wealth, and these social classes to display their newly gained status through the building of specific types of residences or commercial buildings. Conversely, urban change activated new spaces and channels for the discussion and circulation of ideas and the formation of communities engaging with science and technology. Science was indeed fully integral to the geographical transformation of Barcelona, with the 1888 Universal Exhibition being responsible for changes particularly in the open spaces of the city.[30] In the following two sections, I discuss how the Barcelona *gabinetes*—sometimes for reasons beyond their control, sometimes out of disinterest or lack of knowledge on the part of their owners—failed to integrate what they did and were within these two sets of discourses, and I advance some reasons why this might have been the case.

The redesign of Barcelona: *Gabinetes* left behind

Barcelona's transformation into an international metropolis was heavily driven by its bourgeoisie, keen to overcome the city's status as a provincial capital[31] and to assert itself as both Catalan and international.[32] The transformation, which was both physical and cultural, started with the gradual demolition of the city's walls, in stages between 1854 and 1873.[33] The demolition was regarded at the time as transformative moment, with the municipality of Barcelona setting up a *comisión de corporaciones* in 1853, gathering together some of the most privileged and influential groups in the city, with the idea of steering the transformation.[34] The following years saw the middle and upper classes taking the lead, with the municipality and the administration often showing limited capacity in this respect.[35]

Partly following municipal architect Ildefonso Cerdá's 1859 project for a non-hierarchized new city,[36] Barcelona's development took mostly the form of an expansion of the old city toward the Northern neighborhood of Gràcia, which came to be seen as the natural habitat of the newly developing urban upper and middle classes.[37] The urbanization of Gràcia and the surrounding space, L'Eixample (literally, the broadening) started timidly around 1860[38] and accelerated between 1870 and 1885,[39] with the new bourgeoisie building palaces and quality flats in the newly urbanized space. The old city still kept some of its prestige and character—as most shops, banks, and bourgeois social venues were still there[40]—but it was increasingly under scrutiny by the bourgeoisie, who found it was poorly laid out and inconvenient.[41]

During the 1890s, the prestige of the Eixample as the new bourgeois center was further consolidated.[42] Intellectuals, artists, liberal professionals, architects, and engineers joined the bourgeoisie in the new neighborhoods, and this resulted in differentiated, hierarchized areas within the Eixample.[43] At the same time, the area was developing as a commercial hub;[44] with this, the geographical and social center of Barcelona started to inexorably move toward the North: the Plaça Catalunya, initially an empty esplanade between the new and the old city at the North of La Rambla (the main artery of the old city), gained importance during the 1890s[45] and was finally urbanized in 1902,[46] thereby becoming consolidated as the center of the new Barcelona. From 1907 onward, drawing upon the spatial and visual transformations the city had undergone during the previous decades, the cultural movement known as *noucentisme* consolidated the notion of the city as the natural space of modernity—that is, a laboratory in which the cultural and intellectual elites lived and worked in an architecturally monumental background, and were shaped by and in turn shaped it in a climate of constant experimentation.[47] It was also during the decade 1900–1910—that is, almost after the short-lived era of the *gabinetes*—that the narrative of Barcelona's development as a bourgeois victory in imposing progress and modernity started to come into question; social tensions exploded and working-class Barcelona became more neatly differentiated from the bourgeois city in both its physiognomy and understanding of space.[48]

The previously described transformations saw Barcelona being recognized as a major European capital, transformed because of its industrial prowess and bourgeois vitality; the new identity of the city, and of Catalonia more generally, capitalized to a great extent on being perceived as the more European, modern alternative to Madrid.[49] The *gabinetes* and their owners likely benefitted from this expansion and the accompanying climate to a certain extent. Those *gabinete* owners who worked as opticians were, as liberal, educated professionals, part of the Barcelona bourgeoisie—although they were not necessarily among their most prosperous or visible echelons—and others came to phonography from a similarly bourgeois background. For example, Juan B. Baró, the partner of Rosillo at the Sociedad Artístico-Fonográfica, was described in *La Vanguardia* in 1892 as an "intelligent industrialist"[50] preoccupied with the latest developments in aluminum and alloys, and in the same year he applied for, and obtained, permission from the municipality to build two balconies in his house in 329 Consell the Cent.[51] The building was not far from the Passeig de Gràcia and hence part of the Northern expansion of Barcelona, which suggests that Baró indeed belonged to the new bourgeoisie who could afford and was keen to move outside the center.

Nevertheless, even though some individual owners of *gabinetes* might have personally benefitted from the bourgeois expansion, in the newly configured Barcelona the *gabinetes* as a group did not fare particularly well. A key reason for this might have been their physical location. The *gabinetes* were clustered around La Rambla, with four of them (including Grandes Almacenes El Siglo) being on the boulevard itself. Of the other four, no one was more than three hundred yards away from La Rambla, including Sociedad Artístico-Fonográfica, which was the closest to the Eixample and not far away from the university either—a developing area at the time too, attracting mostly a class of academics and intellectuals.[52] As the city expanded, some parts of La Rambla were redeveloped too, and the building where the Roselló *gabinete* was located, in Rambla d'Estudis, was rebuilt rather luxuriously in 1878.[53] Nevertheless, by the time the *gabinetes* started to open, the Rambla and surrounding areas had mostly lost its commercial appeal to the developing bourgeoisie, which would be the natural audience to which the owners of the *gabinetes* could sell their products; the bourgeoisie was instead attracted to the new establishments opening in the Eixample.

Moreover, the fact that the bourgeois urban development was heavily associated to the visual rather than the auditory might have diminished the possibilities of recording technologies. Resina indeed states that, in pushing urban development, the Barcelona bourgeoisie had "visual needs": as an emerging social class, it needed a new space that was sufficiently distinct and identified the bourgeoisie itself as well as its values.[54] A similar notion is advanced by García Espuche when discussing photography: the technology was instrumental not only in documenting urban change in Barcelona, but also in terms of giving significance to certain spaces (i.e., the most photographed) over others, thus contributing to shape the new space and its conceptualizations in profound ways.[55] A further example of how central the visual was in the development of Barcelona is the significance of the decorative arts under *modernisme*, with internal and external furniture and decorations (from tables and chairs to mail boxes, balusters, and plaques) being used to create a sense of visual identity among the Barcelona bourgeoisie. Sound, however, was granted less space in such discourses, which might have similarly disadvantaged the phonograph and prevented it from being elevated at this stage to the status of symbol of the developing bourgeois modernity.

There is a further particularity of Barcelona's urban development that might have proved difficult to overcome for the *gabinetes*: whereas in Madrid many *zarzuela* theaters and *gabinetes* were in close proximity, stimulating the latter and developing new ways of listening to recordings which were intrinsically linked to live music (as will be discussed in Chapter 6), the same did not apply to Barcelona. Theatrical spaces in Barcelona were indeed rapidly evolving in these years too, partly as a result of the overall development of the city, but also in response to shifts in theatrical genres, practices, and audiences. These developments, however, did not turn out to be particularly beneficial for the *gabinetes*, and the *gabinetes*, in turn, were not particularly active in turning these developments into opportunities.

One such significant development was the fusion, from 1890, of theatrical *modernisme* and *teatre catalá*. *Modernisme* had been initially cultivated by a minority of innovative artists and intellectuals seeking to renovate Catalan theater in line with currents elsewhere in Europe, but being mostly interested in achieving commercial success. *Teatre catalá*, on the other hand, had introduced Catalan in the theaters, as opposed to the then-predominant Spanish, and was mostly directed to the nascent local bourgeoisie seeking to achieve cultural hegemony throughout the territory of Catalonia. The new genres resulting from this fusion combined the aesthetic ambitions of *modernisme* with the nation-building aspirations of *teatre catalá*. They grew in popularity throughout the 1890s, and found a space in the old city of Barcelona and the nearby areas of the Eixample.[56] These theaters were just a few hundred yards north from the area where most *gabinetes* were located, but it is likely that the audiences of this new theatrical form, which was based on the spoken word and not on music, would not have found in musical recordings the memento value that these had in Madrid for audiences of *zarzuela* and opera theaters. Nor is there any evidence of the *gabinetes* recording Catalan-language song (both traditional and newly composed), which allegedly would have been popular with the bourgeois audiences of *modernisme* and *teatre catalá*.

Like Madrid, Barcelona too had an infrastructure of lighter musical theater genres (*pantomima*, melodrama, *zarzuela*, *opereta*, vaudeville, *revista*). These thrived between 1895 and 1905 in theaters located in the Avinguda del Paral•lel.[57] The Paral•lel, however, starting south of the Rambla and advancing toward the West, was hardly located in an advantageous position for the *gabinetes* to sell their recordings to theater-goers, as likely happened in some areas of Madrid.

Whereas urban geography might not have benefitted the *gabinetes* particularly, there is also little evidence that their owners took the initiative to promote their products as a memento or a complement to the live music experience even in those cases in which proximity allowed them to do so. Indeed, a very significant opera theater stood in close proximity of the *gabinetes*, with at least five of them located less than three hundred yards from it: this was the Teatre del Liceu, which was also prominent sociability center for the bourgeoisie, the obvious customer base for domestic phonography. But evidence that the *gabinetes* took advantage of this is limited and mixed. For example, with the Liceu having been a well-known Wagnerian hub from 1885, we might have expected the *gabinetes* to take advantage of that and record Wagner rather extensively, but there is not much evidence that they did so: only five recordings of Wagner music made in Barcelona have survived (three by El fonógrafo, one by Moreno Cases, and one by Sociedad Artístico-Fonográfica). Moreover, Ángel Constantí, who features in both the Moreno Cases and the Sociedad Artístico-Fonográfica recordings (the ones by El fonógrafo did not include the names of the singers, as was the norm with this company), was not known as a Wagnerian tenor, specializing instead in *zarzuela*: we might imagine that this probably did not provide much of an enticement to Liceu regulars to acquire his recordings. In June 1899, Sociedad Artístico-Fonográfica was advertising recordings *La Bohème*[58]—which is a rarity, since this *gabinete*, as is the case with most in Barcelona, rarely named specific operas or *zarzuelas* in its press advertisements. *La Bohème* was premiered at the Liceu in April 1898 to great success, and so we might imagine that that Rosillo and Baró, by naming the opera specifically in their advertisements, were trying to attract customers among Liceu patrons who would have seen and enjoyed some of its first performances. It is difficult to gauge whether the *gabinete* was successful in doing this, as they closed shortly thereafter.

Another indication that the Barcelona *gabinetes* lacked sustained connections to the theatrical world is the fact that they did not build strong relationships with singers in the way that the Madrid or Valencia ones did. Indeed, whereas the first advertisements of Madrid *gabinetes* (from 1897) did not normally name singers and instead focused on the phonographs themselves and their accessories— suggesting that the main appeal at this stage was for customers to reassure themselves of the capabilities of the phonograph by producing their own home recordings—by 1899–1900 singers were central to these *gabinetes*' publicity. Madrid and Valencia catalogues from these years were built around individual singers (as opposed to composers, pieces, or genres). They listed the numbers that a particular singer had recorded and often included photographs and carefully chosen biographical details, suggesting that singers now played an increasingly important role in customers' decisions, no longer simply fascinated by the possibility of trying out the phonograph themselves, but instead developing a taste for specific genres, pieces, and voices. There is no evidence, on the other hand, that the Barcelona *gabinetes* issued their own catalogues, and mentioning singers by name in advertisements was rare. A notable exception is Roselló: although no cylinders from this *gabinete* have survived, in an advertisement the owner claimed to sell a wide range of recordings for all sorts of instruments and ensembles and even mentioned one performer by name: José Gomis, a tenor who had achieved modest local success in Barcelona.[59]

More tellingly, at least one of the *gabinetes*, El fonógrafo, did not even label its cylinder cases with the names of performers or announced them at the beginning of the recording. This is true for all ten surviving cylinders from this *gabinete*, which makes it plausible that it was standard practice.[60] The absence of names suggests that the concept of the recording as a commodity of aesthetic value might not have been fully developed at the time in Barcelona, and that the main appeal for customers was still the fact that the phonograph was able to reproduce sound; the exact sound that was reproduced was of less importance. While the earliest evidence from Madrid suggests a similar ontology of recordings, Madrid *gabinetes* then evolved to an understanding of recordings which is more akin to what we have today and to what was happening elsewhere, particularly in the United States; the Barcelona industry, however, failed to develop its understandings of recordings and convey it to customers to the same extent, which provides a fascinating and in many ways unique picture on how ontologies of recordings were evolving at this time—quicker in some places, more slowly in others.

Still, the Barcelona *gabinetes* were faced with some of the same decisions and constraints as those in Madrid concerning how to select singers for recordings, and it is Corrons's cylinders that offer us the most comprehensive insights in this regard. Like most Barcelona *gabinetes*, Corrons did not include the names of singers in his advertisements either. Evidence suggests that the *gabinete* was not always consistent either when it came to naming performers in the cylinder boxes or in the spoken announcements that preceded performances. Five V. Corrons cylinders (out of 25 surviving) do not indicate the name of the performer—but, when it comes to those issued in later years by J. Corrons, it is nine out of 18 which are anonymous, which suggests that the *gabinete* did not start systematically including the names of the performers as the years went by, as was the case with some establishments in Madrid.

Even though the proportion of named cylinders is comparatively small, they still allow us to draw some conclusions. We know, for example, that Corrons employed two singers of some renown: opera tenor Ángel Constantí, rather well known in the Barcelona *zarzuela* scene at the time, and *zarzuela* soprano Blanca del Carmen, one of the first Spanish singers to have a recording career of sorts: she recorded for Viuda de Aramburu in Madrid and became one of the

first Spanish singers to record for Gramophone when they visited the country
for the first time in 1899. Nevertheless, the other singers whose voices have survived in Corrons's cylinders are more obscure, and were no doubt part of the
multitude of working singers of no particular distinction who worked at this
time the Barcelona theaters: a señorita Fernández (perhaps Juanita Fernández),
a señorita Martínez (perhaps Salud Martínez, who was active in Barcelona since
at least 1896, or Pura Martínez, who also recorded for Gramophone), tenor José
Moratilla, a señor Monrás or Monrós. Others, such as señora Pianchini and
señorita Vallrossoll, do not appear to have had a stage career, which suggests they
might have been amateurs or advanced students. They were certainly not an exception in the Spanish industry, and a number of similar cases will be discussed
in Chapter 6.

The Madrid and Valencia *gabinetes*, particularly the larger ones such as
Hugens y Acosta and Blas Cuesta, typically employed a range of performers
of different abilities and reputations. Performers usually (but not always) recorded within their preferred repertoire; recordings could be sold at rather
contrasting price points, depending on the genre of the cylinder and the reputation of the singer. This presumably allowed these *gabinetes* to reach different
pockets of customers—from those who could afford to pay thirty or sixty *pesetas*
for a recording of the newest fashionable opera singer, to those who contented
themselves with a *zarzuela* or flamenco recording costing five or six. Evidence
suggests, however, that the Barcelona *gabinetes*—not even Corrons—never managed to implement a sophisticated pricing strategy, and never hired celebrities on
a par with those who regularly recorded for Hugens y Acosta. Most singers we
know to have recorded for Corrons were relatively unknown: even Constantí or
Del Carmen cannot be said to ever have reached the level of fame of Julián Biel,
Lucrecia Arana, or Leocadia Alba, who all recorded for Madrid *gabinetes*.

Science: Focusing on the practical

A further reason why the Barcelona *gabinetes* were less successful than those in
Madrid and Valencia has to do with their comparatively reduced ability to place
themselves within existing discourses around science and technology, although
such issues were as important in the new Barcelona and in Catalonia as they
were in Madrid. Throughout the last decades of the nineteenth century, as the
city evolved and Catalan nationalism developed, different political factions and
social classes promoted contrasting discourses around science,[61] which crystallized to a great extent in the 1888 Universal Exhibition held in the city. Elite
science practiced at an increasing number of academic and collectively funded
research institutions existed alongside a vibrant public and civic science scene,

with new formats, media, and technologies entering the urban stage from the mid-to-late nineteenth century. In this context, the 1888 Exhibition provided an unparalleled opportunity to experiment with how to make science accessible to a range of audiences.[62] Discourses on science were also intrinsically connected to discourses about Catalan national identity.[63] In the last years of the century, the notion developed that progress and modernity were not simply the natural outcome of science and technology, but rather ends that Catalan society had to actively pursue through a carefully controlled national policy.[64] Even though real self-government was scarce at this time, such a policy was progressively implemented through institutions set up and maintained by civil society, such as the Institut d'Estudis Catalans (founded 1907) and the Mancomunitat de Catalunya (funded 1914). Both institutions were committed to dedicating funds and efforts to train scientists and other qualified professionals, which were seen as key to the development of Catalonia.[65] We see a trace of the broader *regeneracionista* current here: indeed, Catalan nationalism or *catalanisme* can be understood as a regional manifestation of *regeneracionismo*, but it was not homogeneous either.[66]

Although often neglected by existing research,[67] one key group engaging with and generating discourses of science, modernization, and national identity were small business owners and liberal professionals in the areas of applied science and technology—the same class *gabinete* owners belonged to. Numerous such professionals took part in the 1888 Universal Exhibition—including several photographs[68]—and one of the main aims of the exhibition itself was indeed to draw attention to the role of small businesses in the industrial development of Barcelona and Catalonia. From the 1890s onward, many such small businesses and liberal professionals became the main driving force behind the tourism guides that proliferated in those years in Barcelona, aimed at attracting domestic and foreign tourists as well as investors. Small businesses financed these guides through paid advertisements that often were sumptuously illustrated and contained numerous claims about the scientific credentials of the products they advertised. They therefore also contributed to presenting Barcelona as a city that was both industrial and industrious, with a solid petite bourgeoisie or commercial class who had made its fortune through conscientious application and development of technological and scientific innovations from abroad.[69]

Even though tourism guides would have been an obvious forum for the Barcelona *gabinetes* to promote themselves (as it was for their Madrid counterparts), they did not do so. They did not contribute either to publications which voiced the interests of the merchant and small entrepreneur class, and which as such could have helped them frame and develop a discourse on recording technologies, such as the magazine *El fomento: eco del capital, de la industria y comercio*. One reason for this might simply be the personalities and priorities of *gabinete* owners. Whereas several of them were already active in 1888 as opticians or in other science-related professions, none of them took part at the Universal Exhibition, suggesting that they were never particularly interested in acting as leaders within their peer group. Other reasons for the *gabinetes'* lack of success in this respect might have to do with the particularities of the discourses around science and technology developed by shop owners and small businesses at that time—as in the previously mentioned publication *El fomento*. In these, science and technology were almost always framed in terms of practicality and contribution to the developing Catalan economy—particularly in the textile and related areas, which were to a great extent driving the development of the region. Sound technologies, on the other hand, did not have any obvious practical application in the Catalan industry, and this might have been another reason why they went relatively unnoticed. For example, when a delegation of workers of the cotton industry was sent to Chicago in 1893 with a subsidy from the provincial government to visit the Exhibition there, their rather detailed report did not include any mention of the phonographs that were indeed exhibited in Chicago.[70] Overall, while the Madrid and Valencia *gabinetes* were keen not only to boast about their inventions and patents, but also to connect them to broader narratives about how science and technology were regenerating Spain, the Barcelona ones, even though they had no shortage of technological achievements to their credit, failed in doing the latter, perhaps because of a lack of suitable referents in a scientific culture dominated by the visual, on the one hand, and the practical, on the other.

The aftermath of the Barcelona *gabinetes*

In the decades following the era of the *gabinetes*, Barcelona developed a reputation as an important city in the international recording industry—ahead of Madrid—and many of its inhabitants also contributed to shaping a culture of record collecting whose initial years will be discussed in Chapter 7. In parallel, the piano roll industry thrived in the city from 1905 to 1930 thanks to the factory Rollos Victoria—one of the most important and active worldwide, committed to recorded Catalan art music as well as more canonic works. However, the *gabinetes* are not normally discussed as part of the collector-authored narratives of Barcelona as a capital of the international recording industry.[71] A reason for this is that the *gabinetes*, with their cylinders being relatively inaccessible in private collections, were not very well known or easy to learn about until recently, after the Biblioteca Nacional de España digitized their cylinders and other libraries and archives similarly gave the very earliest of recordings greater importance that they had thus far.

But, besides the difficulties to actually find out information about the *gabinetes*, other reasons might have contributed to this silence. Firstly, as explained previously, it is not easy to fit the modestly successful *gabinetes* into these narratives as full-fledged pioneers or astute entrepreneurs. Secondly, it is not easy either to fully ascertain the role that they might have had in shaping what came next. Indeed, whereas in Madrid some of the *gabinetes* survived for years under the form of resellers and representatives of multinationals, this was not the case for Barcelona, and there is no evidence that the individuals involved in the wax cylinder industry then applied their skills to selling gramophones or manufacturing piano rolls. There is no clear continuity, however, between the *gabinetes*, on the one hand, and the gramophone and piano roll industry in Barcelona, on the other, although the latter clearly inherited many of the cultural signifiers and practices first developed by the former.[72] More abstractly, though, we can point at how the sound technology industry developing in the city in the first decades of the twentieth century was as influenced by its physical milieu as the *gabinetes* were—although these influences manifested themselves in radically different ways.

The first multinational to arrive in Barcelona, Gramophone, did so in the autumn 1899 when the company's London office sent a team of engineers headed by William Sinkler Darby to record local artists for its catalogue.[73] The team had already been in Madrid the year before. Gramophone initially planned to open a branch in Barcelona in April 1901,[74] but—even though the company kept visiting Madrid and Barcelona regularly—the opening was delayed to 1903, with the office being located in the corner of Carrer Pelai and the Rambles. The headquarters was north of the previous area of influence of the *gabinetes*, and therefore closer to the developing bourgeois center. Torrent i Marqués claims that the director of the new office was a man of French and German ancestry called Albert Reich or Reig, who is not known to have had connections with the *gabinetes*.[75] A second expanding company, Odeon, opened its own branch in Barcelona in 1906,[76] and was later on acquired by Lindström, another of the big recording consortia of these early years.[77]

The Gramophone and Odeon Barcelona offices imported talking machines and discs from the mother company for distribution in Spain through a number of agents scattered around the country. They also organized recording sessions with local artists, with a preference for indigenous genres and practices. In fact, whereas the *gabinetes* were never very active in recording specifically Catalan or Barcelona genres and styles (e.g., Catalan-language song), the multinationals did—which was consonant with their policies elsewhere in the world. Instead, multinational companies focused on recording indigenous repertoires in each of the countries they established a presence in. This had a two-fold aim. Firstly, it gave companies the ability to cater to the tastes of local or language-specific markets (such as Czech-language recordings made in Europe then being sold among Czech emigrant communities in the United States[78]). Secondly, it enabled multinationals to diversity their global catalogues: indeed, a small percentage of recordings of vernacular and indigenous repertoires were marketed to international audiences with an appetite for exotic recordings.[79]

A prominent example of how these types of practices were deployed in Barcelona concerns the recordings of the Orfeó Català—a choral society who came to be well known from the late nineteenth century onward for their dedication to Catalan traditional and art music alongside choral staples of the Western canonic repertoire performed to high standards. Gramophone recorded the Orfeó Català, with teacher Emerenciana Wehrle as a soloist, in their first visit to Barcelona, and continued to record the choral society regularly over the following years. Eventually, a full series of Orfeó Català recordings was released in 1916, consisting mostly of Catalan-language traditional and art music pieces.[80] To promote the recordings, Gramophone launched a Catalan-language catalogue providing a history of the Orfeó and highlighting its role in the revival of Catalan culture.[81] With the catalogue being in the Catalan language, we might presume that the recordings were intended primarily for domestic consumption—but at the same time the point was made in the text itself that Gramophone had made it possible for the whole world to listen to the Orfeó. The Orfeó, indeed, kept recording regularly during the following years, venturing into standard (i.e., non-"ethnic") repertoire and recording Beethoven's *Missa Solemnis* in 1927 for Victor.[82]

But multinationals did not limit themselves to recording whatever local genres or practices were already in existence. In some cases, they contributed to shaping those genres and practices decisively, as was the case with *sardana*, the traditional instrumental Catalan genre which accompanies a circle dance. At the turn of the century, *sardana* was still slowly expanding outside its original birthplace, the region of Empordà, and establishing itself as a Catalan genre. Whereas there is no evidence that the *gabinetes* ever recorded *sardanas*, Gramophone engineers did so in their first visit to Barcelona in 1900, with the wind band of the Casa Provincial de Caridad orphanage recording five *sardanas* (out of a total of 13 pieces that the band recorded for Gramophone). This was, according to Torrent i Marqués, unusual: first of all, the *sardana* normally requires their own combination of traditional instruments different from a wind band, so, given that these recordings have not survived to our days, it is not clear which instruments were used and what sorts of arrangements were played. Secondly, the *sardana* was at the time not popular in Barcelona, but just in the region of Empordà, which leads Torrent i Marqués to suggest that the band of the Casa Provincial de Caridad might have employed a significant number of musicians from Empordà who would have been familiar with the genre and contributed to importing it to Barcelona.[83] In the next few years, as the genre expanded in Barcelona and Catalonia, other multinationals recorded *sardanas*: Odeon recorded 18 numbers of the Ampordanesa La Principal de La Bisbal in 1905, and it is likely that Pathé made some *sardana* recordings as well.[84] Zonophone, a branch of Gramophone specializing in low-cost records, also recorded *sardanas* in 1906 and 1907.[85] Moving *sardana* recordings from Gramophone to Zonophone was, according to Torrent i Marquès, a business decision having to do with the fact that *sardana* was not popular with the upper echelons of society who were able to pay high prices for recordings; Zonophone's price range, on the other hand, was more affordable to lower-middle and working classes.[86] Another strategy pursued by Zonophone to lower costs included issuing successive re-editions of the same recordings to avoid having to hire performers for a second time.[87] Recording companies thus did not just profit from the surge of interest in *sardana*: but they also contributed to it by documenting its initial years and creating a repertoire of recordings.[88]

Although multinationals shaped to a great extent what and how recordings were consumed, a second wave of local entrepreneurs also contributed considerably to shaping Barcelona as a recording capital—and they did so more decisively than the original Barcelona *gabinetes*. The new entrepreneurs did not produce their own recordings, but they left their own, local trademark on sound technologies and practices in other ways. For example, New Phono, initially founded in 1834 as a music warehouse, built and sold for a short period of time in the decade 1900–1910 their own sound playback devise: the so-called gramoófonos—very similar to gramophones, but with a slightly different name to avoid copyright battles.[89] In 1914 it was no longer selling those, and was instead operative as a re-seller of Pathé recordings.[90]

Other shops and factories in Barcelona keenly promoted themselves as manufacturers of gramophones and talking machines—but what they did in reality was to import the mechanisms from Germany or Switzerland and limited themselves to building the decorative elements of the gramophones (horn and base). These establishments had close equivalents in many cities around the world: for example, in Colombia, local commercial agencies and representatives exported raw materials and imported manufactures which, crucially, did not only include phonographs and gramophones, but also a list of other products (textiles, perfumes, jewelry) that contributed to disseminating a particular idea of urban modernity where global outlooks mixed with local networks and practices.[91]

One of these manufacturers was Arpí A.M., a subsidiary of the music warehouse Serrano y Arpí. In the 1910s, Arpí A.M. presented itself in its advertisements and the labels it appended to its gramophones as "the only national factory of TALKING MACHINES [sic] and all of its accessories, which has acquired an international reputation." At least one gramophone survives

carrying both Arpí A.M. and José Navarro labels, suggesting that Arpí manufactured it and Navarro distributed those in Madrid.[92] Arpí A.M. were more concerned with establishing their own prestige and building an identity consonant
with discourses around science and modernization than the city's *gabinetes* had
been, and in 1914 they attended the Exposición y Concurso Internacional de
Barcelona, obtaining an award which they publicized in their advertisements
and disc cases.

A similar company, Compañía Española de Máquinas Parlantes, was founded
by Eusebio Vicente around 1906 or 1907.[93] Eusebio died in 1908 and his son César
took over, moving the company from Fontanella 10 near the Plaça Catalunya to
the more bourgeois Passeig de Gràcia—the same area in which New Phono was
located. The Compañía issued its own catalogues, containing recordings from
labels such as Odeón and Fonotipia and focusing mostly on *zarzuela*, opera, and
a few token examples of other genres. They also issued disc cases with their own
moderniste-style logo representing a siren, and their own gramophones ranged
from the rather modest, with only very minimalist decoration, to the more luxurious.[94] Toward the mid-1910s, the Compañía opened a branch in Madrid, and
it remained influential in Barcelona for decades after that: in the 1930s it was one
of the forces behind Discófils, a group of record collectors formed by a group of
engineers, intellectuals, and artists coming mostly from the left-wing Barcelona
bourgeoisie. Phono-Odeon also manufactured gramophones throughout the
1910s; it is not likely that they had any connection with recording label Odeon,
but instead just used the name for prestige reasons.[95] Finally, from Carrer Pelaig,
and hence between the new and the old city, Velten, Puig y Compañía also dabbled in the import and "decoration" of gramophones, producing one in which
the box shows a caricature of a Catalan *pagès* (peasant) wearing a *barretina* (traditional hat)[96]—a further example that local entrepreneurs were still shaping
and adapting imported products and practices for local consumption, with the
second generation of record shops being able to integrate themselves in their milieu more successfully than the *gabinetes* had been able to.

5

Gabinetes fonográficos in Valencia, 1899–1901

Next to Madrid and Barcelona, a further city, Valencia, emerged in the years around 1900 as an important center of the budding Spanish recording industry. Only four *gabinetes* operated in Valencia, but their contributions to the early history of the phonograph in Spain rank alongside those from their counterparts in Madrid. The Valencia *gabinetes* have left us more than 150 cylinders,[1] as well as a publication (the magazine *Boletín fonográfico*) that is key to understanding the logistics and discourses of the early industry; as a primary source, it is unique not only in the Spanish context but also more globally. While it is not possible to conclusively ascertain the reasons why Valencia—and not other large cities such as Zaragoza or Seville—was able to establish itself at the time as a center for recorded music, it is likely that some of the local particularities of the industry as well as the city's spaces had a significant influence. Valencia was, at the time, relatively small and compact compared to Madrid and Barcelona, and was home to just a handful of theaters hosting touring companies. The times when the Valencia *gabinetes* were open overlap to a great extent, unlike in Madrid, where *gabinetes* kept opening and closing between 1896 and 1905. It is likely that the Valencia industry could therefore operate in a more interconnected fashion than their counterparts elsewhere, with *gabinetes* simultaneously competing and collaborating with each other. Building upon Chapters 3 and 4 on Madrid and Barcelona, the Valencia case study illustrates how the early history of recording technologies in Spain was closely connected to place and to specific communities, resulting in highly localized practices that defy generalization. This chapter will discuss how the Valencia *gabinetes fonográficos* operated, how they articulated a discourse around recording technologies that had some distinct features compared to Madrid, and how the nascent interest in recording devices and wax cylinders made an impact and in turn profited from musical life in Valencia.

Gabinetes in the provinces

Before I discuss the Valencia *gabinetes*, I would like to make a brief detour to explore how recording technologies made their way into other Spanish cities

Inventing the Recording. Eva Moreda Rodríguez, Oxford University Press. © Oxford University Press 2021.
DOI: 10.1093/oso/9780197552063.003.0006

and towns at the time. At least a further 18 *gabinetes* existed outside Madrid, Barcelona, and Valencia; information about these is presented in Table 5.1. While Valencia was at the time regarded as part of the provinces, and not a major European city like Madrid and Barcelona, we should in no way assume that Valencia's recording industry is representative of the activities of the other provincial *gabinetes*. Indeed, while the existence of these *gabinetes* suggests that recording technologies were at the time timidly becoming more visible throughout Spain, their activities remained rather precarious. Surviving cylinders are scarce,[2] and written sources about the *gabinetes* are scarce too: some *gabinetes* might not have even advertised in the local press, perhaps relying on word-of-mouth instead.[3] This suggests that provincial *gabinetes* did not produce very extensively and were not as concerned as those in Madrid to advertise their products as part of a discourse on technology, modernization, and national identity.

With some of the *gabinetes* we know about, we cannot even ascertain whether they produced their own material or simply resold others'. With Madrid and Valencia *gabinetes* being especially keen to sell their products in the provinces, it is conceivable that they allied with local businesses in their attempt to do so; these local businesses might or might have not produced their own material on the side. Lacaze Óptico, in Zaragoza, did sell cylinders from other *gabinetes* as well as producing its own,[4] while Emilio Zurita and Societé Edison Eladio Carreño, both in Gijón (Asturias), likely worked as resellers: Spanish collector Carlos Martín Ballester names them in his list of *gabinetes*, but no cylinders or evidence that they indeed produced their own material has survived.[5] Viuda de Ablanedo e Hijo was a *gabinete* in Bilbao; the Ybarra family archive at Eresbil counts more than one hundred cylinders with the Viuda de Ablanedo label on the box. Some of those have been digitized, with the spoken announcement at the beginning of each recording revealing that they were really made by Viuda de Aramburo in Madrid. There remains a sizeable contingent of non-digitized cylinders; some were likely made by Viuda de Aramburo as well, as they feature performers known to have recorded for the Madrid *gabinete* (Blanca del Carmen, Eloísa López-Marán, Rafael Bezares, Señor Navarro, El Mochuelo, Emilio Cabello). Others, featuring Basque music (six *zortzikos* by a señor Maguregui), might have been indeed made by Viuda de Ablanedo in Bilbao.[6]

Unsurprisingly, many *gabinete* owners outside the three largest cities in Spain came from the applied technology industry or had a background in introducing visual spectacle technologies in their cities—such as Manuel Anaya in Béjar, who was also an operator of *cuadros disolventes*,[7] several clockmakers, and José Guerrero Laplaza, a cabinetmaker from Albacete. Guerrero Laplaza filed in 1899 a patent for a wooden horn, but there is no evidence that he developed the invention commercially, as is the case with many phonograph-related patents at this time.[8] Luis Casares, from Casa de Casares in Granada, worked for Pathé's factory in Chatou until 1902, so it is likely that he returned to his native Granada and opened his *gabinete* after that.

Table 5.1 *Gabinetes fonográficos* active outside Madrid and Barcelona

Name	Location	Dates active	Key individuals	Own recordings?	Other activities	Other
Casa de Casares	Granada	1902–1903 (?)	Luis Casares, A. Casares Aceituno	Yes	Mechanic	Luis Casares worked for Pathé until 1902
Casa Erviti	San Sebastián	?	?	No (?)	Musical instruments and sheet music	
Emilio Zurita	Gijón	1901–1903 (?)	Emilio Zurita	No (?)	Industrial equipment	Became a Gramophone representative in 1903
Enrique García	Bilbao	?	Enrique García	Yes	Piano warehouse, café	Became a Gramophone representative in 1904
Francisco Rigau	Tarragona	1899–1903 (?)	Francisco Rigau	Yes (?)	Clockmaker	No cylinders survived, but advertised recordings by local artists; became a Gramophone representative in 1903
Hércules Hermanos	Valencia	1900–1901	José Hércules, Vicente Hércules, Vicente Peydró (music director)	Yes	Drugstore	
Hijos de Blas Cuesta	Valencia	1899–1902	José, Francisco and Federico Cuesta	Yes	Drugstore, pharmacy	Became a Gramophone representative in 1904
Iturrioz	Vitoria	1901	Victor Iturrioz	?	Clockmaker	
J. Viola // Bazar González	Jerez	1900	J. Viola	?	Optician	

Continued

Table 5.1 *Continued*

Name	Location	Dates active	Key individuals	Own recordings?	Other activities	Other
La Oriental	Zaragoza	1897–1899	Emilio Gutiérrez	Yes	Photography, optician	
Lacaze Óptico	Zaragoza	?	Luis (Louis) Lacaze	Yes	Optician	
Manuel Anaya	Béjar	?	Manuel Anaya	Yes	Photography, visual spectacles	
Pallás y Cía.	Valencia	1899–1901	Manuel Pallás, Maximiliano Thous	Yes		
Prudencio Santos Benito	Salamanca	?	Prudencio Santos Benito	Yes	Department store, photography	Became a Gramophone representative in 1903
Puerto y Novella	Valencia	1899–1901	Vicente Gómez Novella, Puerto brothers	Yes	Photography, cinematography	
Societé Edison Eladio Carreño	Gijón	?	Eladio Carreño	No (?)	Drugstore	
unknown	Almería	1899	Fernando Salvador Estrella (?)	Yes (?)		No cylinders survived, but advertised recordings by local artists
Viuda de Ablanedo e Hijo	Bilbao	?	?	No (?)		

The very reduced numbers of surviving cylinders from outside Madrid, Barcelona, and Valencia (no more than six for any *gabinete*) suggest that most or all of these *gabinetes* might have been part-time ventures. The customer base they had access to in their immediate surroundings was more reduced than it would be in a larger city, and they also had to face competition from the most active *gabinetes*, such as Hugens y Acosta and Puerto y Novella, who sold in the provinces either by mail order or through appointed representatives, so it is unlikely that they could sustain their business on a full-time basis. Similarly, they would have access to smaller numbers of musicians they could hire to make recordings.

The surviving cylinders from these *gabinetes*, nevertheless, offer a few insights which complement what we know about the early recording industry in Spain from the *gabinetes* in Madrid, Barcelona, and Valencia. The first of these insights concerns repertoire. It is significant that three (out of four) of the surviving cylinders from La Oriental in Zaragoza (one of the first *gabinetes* to open in Spain, in spring 1897)[9] and all six from Enrique García in Bilbao featured traditional vocal repertoires heavily associated with their regions: *jota* for the former and *zortziko* for the latter.[10] Moreover, all surviving La Oriental cylinders come from Pedro Aznar's collection (based in Barbastro, Aragon, the region of which Zaragoza was the capital), and all of Enrique García's come from the Ybarra family collection, also based in Bilbao, which suggests that some of the provincial *gabinetes* recorded traditional music primarily for their local market and did not try to venture outside.

Nevertheless, other sources suggest that the situation was more complex than that. Flamenco (originally an Andalusian genre) was recorded everywhere, and presumably sold everywhere too, as the genre had by that point expanded well outside its native Andalusia and was well-loved by urban audiences elsewhere. There is no reason to think that the flamenco cylinders recorded in Madrid or Valencia would have been primarily exported to Andalusia: their numbers are too high to presume the Andalusian market could absorb them all, and industry publications sometimes complained that demand for flamenco cylinders among customers was so high that *gabinetes* were forced to produce these instead of focusing on more prestigious genres such as opera.[11] With other traditional genres, the situation is less clear. There are examples of *zortziko* and *jota* recorded outside the Basque Country and Aragon, respectively, particularly in Madrid and Valencia; however, we do not know whether these repertoires would be mostly exported to the regions they originated from or whether they would be sold throughout Spain. The latter proposition, though, seems more likely, for two reasons. Firstly, traditional genres other than flamenco often received a modicum of attention too outside their region of origin; for example, the *jota* was well known outside Aragón thanks to its widespread use in *zarzuela* numbers. Secondly, in their early history, recording technologies often acted as a gateway for traditional repertoires from specific locales to become known outside their place of origin and for the idea of national and transnational repertoires to develop. This is the case with multinational catalogues through the 1900s and 1910s, in which a small percentage of the recordings of vernacular and indigenous repertoires made initially for their community of origin were marketed to international audiences with an appetite for exotic recordings.[12] Michael Denning's book *Noise Uprising* is concerned with how a range of urban genres originating in port cities in the second half of the 1920s combined through the making and circulation of recordings as a "soundscape of modern times" that expanded across the globe.[13] We should not assume that such large-scale developments did occur during the *gabinetes* era, as only a small section of the Spanish population had regular access to recordings at this time. However, it is plausible that the *gabinetes* might have planted the seeds of this by distributing recordings of flamenco, *jota*, and *zortziko* all over Spain. This too resonates with my discussion in Chapter 2 about recording technologies traveling all over Spain, helping shape connections between geographical areas and social classes.

The Valencia *gabinetes*

As with Barcelona, commercial phonography made a somewhat belated appearance in Valencia. The first *gabinete*, Hijos de Blas Cuesta (originally a drugstore), did not become active as such until January 1899. As with other establishments that opened toward the end of the traveling phonograph era, the Cuestas started off by exhibiting a Columbia graphophone in their shop from March 1898; soon, customers started to inquire about acquiring one, as well as recordings, and so the Cuestas decided to start their own *gabinete*.[14] In their initial three months, they allegedly sold one hundred graphophones and phonographs.[15] Manuel Pallás opened his own establishment in April of the same year, and his background was fully consonant with *gabinete* owners in the rest of Spain: he had experience in working for the national railway, the telegraph, the telephone system, and as an electrician.[16] Interestingly, like José Navarro and Armando Hugens in Madrid, he was also a traveling phonograph operator for two years before opening his *gabinete*.[17]

It is likely that it was the opening of Blas Cuesta's *gabinete* that sparked the interest of Pallás and of other local professionals working in the area of applied science and technology. Indeed, accounts indicate that Hércules Hermanos started operation directly under the influence of Hijos de Blas Cuesta at some point in 1899,[18] since Valencians assumed in those early days that drugstores were the places where phonographs and recordings were bought, and started going into the Hércules store to enquire about these types of products.[19] Finally, the fourth Valencia *gabinete*, Puerto y Novella, started operations in October 1899.[20] Its owners also had backgrounds comparable to other *gabinete* operators: the Puerto brothers were shopkeepers (with one of them a lawyer[21]) and their business partner, Vicente Gómez Novella, was a local painter and photography aficionado.[22]

Unlike the other Spanish cities outside Madrid and Barcelona, Valencia managed to become a sizeable center in the nascent Spanish recording industry, partaking in some respects of the discourses around science, modernization, and national identity that the Madrid *gabinetes* had contributed to disseminate, while at the same time showing some unique characteristics and leaving a range of written and sound sources that allow us to explore some of the day-to-day mechanics and practices of the nascent industry in deeper ways than are allowed by the available evidence for Madrid and Barcelona. Several reasons can be cited to explain the success of Valencia in this respect: it was, indeed, the third most populous in Spain at the time (213,550 inhabitants, although still less than half than Madrid and Barcelona), and was enjoying considerable socio-economic development compared to other areas in Spain. The region of Valencia was in 1900 the third largest industrial region of Spain after Catalonia and the Basque Country, with Valencia's main industries being textile, shoemaking, wood, and paper.[23] Agriculture, the traditional foundation of Valencian economy, was thriving too, with numerous exports abroad and even the urban professional classes keeping small farms or lands part-time. [24] Moreover, both the city and the region had benefitted, throughout the last decades of the nineteenth century, from the development of road and railway communications,[25] and a new industrial class had emerged as a result; it is likely that the *gabinetes* drew most of their local customer base from this class.[26]

A second reason for Valencia's prosperity has to be sought, as is the case with Madrid, in the personalities of the *gabinete* owners and others who contributed to the industry in the city (some of whom will be discussed later): indeed, with Valencia being at the time a relatively compact city, geographically and socially, it is understandable that the *gabinetes* there appeared to work in a more coordinated way than in Madrid or Barcelona. This sometimes expressed itself under the form of competition—with *gabinetes* lowering their prices[27] or hiring well-known singers in order to compete with each other—and sometimes under the form of collaboration: this can be seen in the fact that all four *gabinetes* advertised and regularly contributed to *Boletín fonográfico*, which therefore became, as will be explained later, a crucial voice in the defense of the interests of the nascent recording industry in Spain, and particularly that of Valencia.

Boletín fonográfico: Musical life, phonography, and discourse

In its first issue, launched in January 1900, *Boletín fonográfico* announced its aims as follows:

> The keen interest that has been developing for the last two years around the phonograph and that is still growing, and the need that every amateur feels not only to know the technological advances in this wonderful device but also to

have some sort of guidance of what he might need to do in order to feed this beautiful yet useful pastime, have encouraged us to start publishing this *Boletín*, in which, without making any promises, we will try to satisfy our readers as best as we can.[28]

The magazine indeed published a significant amount of information that readers would surely find useful. It offered plenty of tips and advice for producing home recordings,[29] and *gabinetes* from Valencia, Madrid, and Barcelona advertised there with lists of recordings for sale. Further details about available recordings were given in regular features about the *gabinetes* themselves and on individual singers. These lists remain useful in our days: they significantly enhance our understanding of the repertoires and musicians recorded, supplementing what we already know from the surviving cylinders. For example, it is thanks to the magazine that we know that Francisca Segura, a relatively well-known *género chico tiple*, recorded for Madrid *gabinetes*, as none of her cylinders have survived,[30] and the same goes for well-known operatic tenor Francisco Viñas, who will be discussed in more detail later in this chapter.

Beyond information about individual cylinders, though, the insights provided by *Boletín fonográfico* complement and challenge what we know about the recording industry in Madrid and Barcelona in two fundamental areas: firstly, they shed light on several practicalities regarding the day-to-day business of the *gabinetes fonográficos*, from the physiognomy of their premises and the make-up and backgrounds on their staff (allowing us to draw some conclusions, albeit incomplete, on recording practices in the studio at this early stage), to the broader impact that the introduction of commercial phonography might have had on the musical and commercial life of a sizeable, but not capital, city. Secondly, it provides a further source for the study of the discourses that connected recording technologies to modernization and *regeneracionismo*—with some significant differences with respect to Madrid that will be discussed in due course.

Boletín fonográfico and transformations in musical life in Valencia

In its early days, the Spanish recording industry employed a multiplicity of singers and other musicians whose working lives were impacted in some ways by the new technologies, and *Boletín fonográfico* is an invaluable source in providing us with the names and other details about these musicians who remain, for the most part, unknown today. Each issue of *Boletín fonográfico* published several profiles of singers who recorded for the *gabinetes*, giving details about the individual's career and his or her recording activities. This was clearly a publicity tactic aimed at encouraging individuals who might have heard a specific singer on stage to buy his or her cylinders, and so some details (especially exaggerated praise) should not always be taken at face value. Specially interesting here are the profiles of singers—professional, semi-professional, and amateur—active mostly or exclusively in Valencia, such as tenors Lamberto Alonso and Jesús Valiente and *tiple* Amparo Cardenal. Other singers who recorded were in Valencia on a visit to sing at one of the five professional theaters (one specializing in opera and the other four offering *zarzuela*, often in combination with other genres).[31]

Since I deal with the singers who recorded for the *gabinetes* in Valencia and elsewhere in Chapter 6 (including the previously mentioned three individuals), I will not dedicate any further attention to them here. What is particularly interesting and more unusual at *Boletín fonográfico* compared to sources for Madrid and Barcelona, is the fact that this publication also gives us details about other musicians who supported singers and fulfilled a variety of tasks at the *gabinetes*. Among those were, first of all, the accompanists, drawn from the pool of local pianists and organists: José Bellver at Hijos de Blas Cuesta, who also played piano in a café and had written a few successful *zarzuelas*;[32] José María Lluch and Juan Cortés (also a church organist) at Puerto y Novella,[33] and Carmelo Bueso Beltrán at Pallás. Bueso Beltrán started his career in church music, then played piano at a café and finally established himself as a choirmaster for *zarzuela* theaters as well as a singing accompanist and coach.[34] Other musicians, employed presumably on a part-time basis, included Rafael Rodríguez Silvestre, a composer of *zarzuelas* and arranger of wind band music, who conducted wind bands at Hijos de Blas Cuesta;[35] and maestro Goñi, who taught orchestral and chamber music at the conservatoire of Valencia, led a sextet that played in Valencia's cafés and was also employed by Hijos de Blas Cuesta to conduct his own ensemble. The two surviving cylinders by the Sexteto Goñi are among the very few examples of chamber music recorded by the *gabinetes*.[36]

Among the musicians employed by *gabinetes*, one played an especially crucial and intriguing role: the "director artístico" or artistic director. From *Boletín fonográfico*, we know that at least three of the Valencia *gabinetes* had one: at Hijos de Blas Cuesta, Bellver doubled up as accompanist and artistic director, whereas Hércules Hermanos employed Vicente Peydró, a local composer of *zarzuelas*.[37] Pallás, on the other hand, employed a non-musician: Maximiliano Thous, editor-in-chief of the local newspaper *El Correo* and also a playwright; he also recorded spoken word cylinders for Pallás.[38] Finally, José Bayarri, the orchestral conductor at Teatro Tívoli also moonlighted occasionally as an artistic director for some local *gabinetes*, although it is not clear which ones.[39] From *Gabinete fonográfico* we also know that José Navarro in Madrid employed composer Luis Foglietti as artistic director;[40] it is likely that other *gabinetes* there and in Barcelona, at least the most active ones, hired local musicians in a similar capacity, perhaps on a part-time basis.

The duties of the *gabinetes'* artistic directors were not clearly laid out by *Boletín fonográfico*, and neither were the day-to-day routines and dynamics of the recording studio or *salón de impresionar*, but we might hypothesize that artistic directors played at least two crucial roles. One might have to do with the selection of repertoire. *Gabinetes* recorded mostly vocal music, focusing on opera, *género chico*, *zarzuela grande*, and flamenco.[41] Their strategy seems to have been mostly guided by what audiences would have demanded based on their experiences of live music: indeed, the *gabinetes'* catalogues included a considerable number of recent hits, but they were also concerned with recording frequently performed repertoire. In this context, the artistic director would have probably had a role in identifying and suggesting suitable works, drawing on his theatrical experience when this was available. Similarly, those artistic directors who were also employed by or had contacts at local theaters would have had access to a pool of singers to be approached for recording.

The second key role of artistic directors happened mostly during recording sessions. We know that artistic directors tended to be present at those,[42] and, with technological know-how developing rapidly in those years regarding how recording technologies could capture musical sound, we can presume that artistic directors would have been significant in liaising between the *gabinetes'* owners (who would be familiar with the phonograph's technicalities, but did not tend to have a background in music) or other staff employed to operate the phonograph, and the musicians who were being recorded. In this context, artistic directors, being trained or professional musicians familiar with recording technologies, might have helped the former identify, through trial-and-error, what counted as a good impression and what the technical means to obtain these would be. They might have also helped musicians adapt and twist their performance so that it could be captured successfully by the phonograph.

It follows from this that artistic directors likely made a key contribution to shaping to a great extent the nascent aesthetic of recorded sound. This developing aesthetic underpins numerous *Boletín fonográfico* articles, particularly those providing technical advice on how to record, and was still dominated to a great extent by the notion of fidelity to the original. A 1901 article, for example, claimed that "a good recording must be a reflection of the natural sound and possess musical tone and color; inferior items will not have these qualities at all."[43] We might presume that artistic directors would, over time, develop a sense of how to maximize these qualities in recordings.

Besides musicians, crucial to the development of the early recording industry in Valencia were also individuals with a background in applied science and technology, either professionally or as amateurs. This includes, of course, the owners of the *gabinetes* themselves, but also several other individuals active in the phonographic industry, such as Tomás Trénor Palavicino, the marquis of Turia, who developed a diaphragm he called The Keating. Trénor Palavicino himself claimed that he took it to developing this innovation because the Bettinin diaphragm was too heavy to transport and, with import taxes, it could easily cost as much as the phonograph itself.[44] Trénor Palavicino was a former artillery captain active in the cultural and scientific life of his city: in 1909 he organized the regional exhibition of Valencia.[45] José Alcañiz, on the other hand, was a local mechanic who repaired phonographs in his workshop,[46] and Cayetano Fiol Ridaura—a soap maker[47]—started manufacturing blank cylinders in March 1900.[48] Casa Cabedo, a local sheet music shop, also developed their own diaphragm[49] and advertised "novelties for the phonograph" in *Boletín fonográfico*: these were of scores of newly premiered operas, presumably so that they could be recorded at home by the phonograph owner.[50] Where it is not possible to establish exactly how much these individuals and establishments would have benefitted from the introduction of the phonograph in Valencia, their trajectories suggest, again, that the phonograph scene in Valencia worked in a more compact, coordinated fashion than in Madrid and Barcelona—presumably because of the size of the city itself.

A final area where *Boletín fonográfico* proves itself as an invaluable source in terms of providing details about the phonograph's impact on musical life and practices in Valencia concerns the day-to-day routines as well some practical business decisions made by the *gabinetes*. Whereas some of the *gabinetes* in Madrid and Barcelona were said to have started in a rather rudimentary and haphazard way, with a phonograph simply being installed in the back room of an existing shop,[51] *Boletín fonográfico* indicates that Hijos de Blas Cuesta purposefully looked for new premises to open their *gabinete*, finally settling on a property they deemed acoustically suitable.[52] We also learn that Puerto y Novella held recording sessions from 9:30 to 12:30 and from 4:00 to 9:00.[53] This similarly suggests that Puerto y Novella, like Hijos de Blas Cuesta, were sizeable operations that likely employed several people (even if only on a part-time basis) who saw their working lives changed in some measure as a result of the emergence of recording technologies. Nevertheless, it should not be assumed that all *gabinetes* active in Spain at this time were as large or as dedicated as Hijos de Blas Cuesta and Puerto y Novella; together with others such as Hugens y Acosta, José Navarro, and Viuda de Aramburo, these might have been the exception rather than the norm.

Boletín fonográfico: Science, modernization, and local identity

Boletín fonográfico's view of the recording industry in Valencia is clearly a benevolent and optimistic one throughout: one in which the phonograph had substantially transformed Valencia and made it into a leading center of the recording industry in Spain. At one point, the magazine claimed that "hundreds of families" lived off the industry, although it is not clear whether this refers to Valencia only or the whole of Spain;[54] details about the working lives of *gabinetes* staff provided in previous sections also suggest that such individuals worked part-time, and not full-time, in the recording industry. Most blips in the sector were confidently labeled as seasonal: for example, in its first year of operation the magazine stated that the industry had suffered some deceleration over the summer when potential buyers were on holiday[55] and then a notable increase over Christmas, with *gabinetes* in Madrid competing by lowering their prices.[56] Only in its later issues did the magazine concede that the industry in Spain had somewhat stagnated. According to an anonymous writer, this was because phonograph factories were still producing the same models first launched in 1896–1898, without bothering to develop new models that could be sold at cheaper prices, whereas the production of blank wax cylinders had considerably increased thanks to the opening of factories in Europe (including Valencia), but this had not contributed to making impressed cylinders cheaper.[57]

This rather rare less-than-optimistic take might invite us to reconsider the generally celebratory and confident tone of the magazine, as well as its claim that it was independent (it is true, however, that its owner and editor, local journalist Manuel Torres Orive, was not directly employed or had connections to any of the *gabinetes*). Indeed, other sources paint a similarly mixed picture: *El cardo* did name Valencia as the second most important center of the industry after Madrid,[58] and the local newspaper *Las provincias* claimed in April 1899 that within the last year the phonograph had become so popular among all social classes in the city that it was expected that most families would take a phonograph outdoors with them on the occasion of the picnics they would hold during the Easter holidays.[59] Even though all social classes could indeed listen to and operate phonographs at public demonstrations and phonographic salons, as has been discussed in Chapter 2, owning a phonograph was at the time mostly limited to the middle and upper classes, so *Las provincias*'s claims seem exaggerated. On the other hand, local tourism guides of Valencia, as well as the published programs for local *fiestas* (which indeed intended to offer an overview of the business and shopping opportunities in the city, to both locals and visitors) from these years do not include mentions of the *gabinetes*, whereas the equivalent guides in Madrid did, suggesting that their importance within city life might not have been as prominent as suggested by *Boletín fonográfico*. Ultimately, *Boletín fonográfico* must be regarded as a magazine committed to defending the nascent industry's interests, and so its validity as a truthful source must be examined without losing sight of the broader discourse the publication was trying to articulate, which will be discussed subsequently. On the other hand, though, studying this discourse is valuable in itself, as it sheds further light on how the *gabinetes* were positioning themselves within discourses around science, modernization, and identity.

Many of the main features of this discourse coincide with developments in Madrid that have been discussed in Chapter 3. Like Hugens y Acosta and Ureña, Puerto y Novella and Hijos de Blas Cuesta were keen to emphasize that they sold outside their city and had their own networks of agents and representatives,[60] and this was often amplified by the local press, with *Las provincias* claiming that the Valencia *gabinetes* were among the most productive of Spain and exported to all the provinces,[61] even though there is no evidence that Pallás or Hércules Hermanos had representatives outside Valencia, which is consonant with the fact that only minimal numbers of cylinders from these two establishments have survived compared to Hijos de Blas Cuesta and Puerto y Novella.[62]

As was the case with *gabinetes* in Madrid, *Boletín fonográfico* also emphasized the aura of practical, sensible luxury which surrounded the establishments' premises in Valencia. The premises of Puerto y Novella were described in the following tones:

A large space, lavishly decorated, with top quality artistic enhancements; a room for the customers, with highly valuable antiques, rugs, amphoras, china, chests and a multiplicity of objects that give the room a serious, elegant appearance, uncommon in these kinds of establishments.[63]

Perhaps most significantly, *Boletín fonográfico*, being more lavishly illustrated than *El cardo* and general newspapers, has also provided us with photographs of some of the *gabinetes'* premises that act as invaluable sources to reconstruct the physiognomy of these establishments. For example, the photographs of Pallás show both the *salón de impresiones* and the sitting room for customers.[64] The former appears rather crammed, with a piano elevated on a dais (likely to achieve an improved balance between the singing and the accompaniment) and, on the other side of the room, a number of phonographs, horns piled on top of each other, a small cabinet with cylinders, a board on the wall with photographs and cards (perhaps from the singers they recorded, as these usually produced and sold cards of themselves as a way of self-promotion) and a small shelf on a corner with a number of what seems to be alcoholic drinks and spirits.[65] The room where customers tried out phonographs and cylinders is more spacious, with a tiled floor, bentwood chairs on the sides, and cupboards containing recordings—a sober environment reminiscent of spaces of masculine sociability. By contrast, Puerto y Novella's listening salon was considerably more lavish, with armchairs, a coffee table, and numerous vases and mirrors.[66] This *gabinete* possessed no less than three *salones de impresionar*: one for music with piano accompaniment, with a second, gigantic horn connecting the phonograph to the back of the piano to improve sound capture; one for military bands; and one for flamenco with guitar accompaniment.[67] It can be presumed that the decision to have three separate rooms had to do with the different acoustic conditions required by different combinations of instruments; at the same time, from the point of view of customers, it would be part of the broader impression of luxury and opulence cultivated by the *gabinetes*.

As with *gabinetes* in Madrid, luxury was understood here in terms of reasonable, practical luxury and closely connected to science. All of Valencia's *gabinetes* developed their own additions and improvements to the phonograph, and these were regularly featured in *Boletín fonográfico*'s articles and in the *gabinetes*' own advertisements. Puerto y Novella invented a diaphragm whose effects were described in *Boletín fonográfico* in the following terms:

> The sound is not only as loud as the original is, but it also comes out of the horn with perfect cleanliness, and it produces, especially on the high notes, such a surprising effect that someone who was not seeing the phonograph would think they are listening to the human voice or the orchestra.[68]

Puerto y Novella also claimed to have invented extra-long cylinders measuring 11 centimeters. These allegedly allowed phonograph operators and private customers to record pieces of music longer than the three minutes allowed by regular cylinders.[69] This would have indeed been a significant breakthrough, not only in Spain, but worldwide, where extending the playing length of cylinders was a pressing matter for the industry too. However, none of these extra-long cylinders have survived, and *Boletín fonográfico*'s article is limited to a very brief article without technical details, so it is plausible that the announcement was intended as a publicity stunt rather than anything else. Puerto y Novella might have indeed been able to manufacture a few extra-long cylinders in their workshop, but this is not to say that they would have been viable from a commercial point of view. The famed Valencian novelist and journalist Vicente Blasco Ibáñez claimed that Puerto y Novella was the only *gabinete* to persuade famous tenor Francisco Viñas to record for the phonograph, since their recording processes were far superior to any other's.[70]

The Valencia *gabinetes* were particularly concerned with two technological challenges. One of them concerned the issue of suppressing the imbalance that appeared whenever a solo voice and a piano were recorded together. This, indeed, was a widespread concern worldwide in the early years of recording technologies, since vocal recordings were the most popular and therefore companies had an incentive to make sure they were made at the highest possible standard; recordists often tried to solve the issue by positioning the singer relative to the piano in creative ways.[71] Instead, Pallás invented a diaphragm which was said to address this issue, as well as improving the "clarity, sonority, and sweetness" of the cylinders.[72] Hijos de Blas Cuesta also claimed in its advertisements to have introduced "improvements" to the recording process to solve this same issue.[73] It is not clear what these improvements were, but, going by what was usual practice in the global recording industry those days, it could be that Blas Cuesta was simply experimenting with positioning the singers and piano in different ways so as to optimize the sound signal.

The second challenge Valencia *gabinetes* were occupied with concerned the viability of manufacturing their own recording and playback devices, as opposed to having to import them from the United States. This was indeed a recurring debate in the Spanish early recording industry, as *gabinete* owners felt that the import taxes for phonographs were extortionate and disadvantageous for them; manufacturing devices in the country would allegedly solve the issue and allow them to sell devices at cheaper prices.[74] Petitions from the *gabinetes* to the Spanish tax office (*Hacienda*) to reduce import taxes were not successful.[75] In response to this, Puerto y Novella announced in August 1900 that they had started producing their own graphophones.[76] It is likely, however, that this was again intended for publicity purposes and that not many graphophones were manufactured or sold. Subsequent issues of *Boletín fonográfico* did not mention Valencia-manufactured graphophones again, and a year later another article in the magazine complained that Edison, Pathé, and a few others still had a monopoly in the manufacturing of devices,[77] suggesting that Puerto y Novella's initiative was not successful.

Even though the discourse around science, modernization, and recording technologies cultivated by the Valencia *gabinetes* through *Boletín fonográfico* is certainly consonant with that which we find in Madrid, a sense of national pride or identity is not as obvious in the former as it is in the latter—although certainly all of the elements discussed so far (trade across the provinces, luxury, science, and technology) had connections to the *regeneracionista* discourse the Madrid *gabinetes* subscribed to. Instead of national rebuilding, what we find in Valencia is more akin to a sense of local pride, which sometimes transformed into marked competition with Madrid. This is fully consonant with regional rivalries and tensions at the time, with elites in the provinces frequently criticizing Madrid's *centralismo*. Even though such rivalries and tensions were generally more common in regions with a strong sense of national identity such as the Basque Country and Catalonia, the reason why Valencian *gabinetes* took the initiative here above their Bilbao and Barcelona counterparts had likely to do with the fact that they were more successful commercially and more coordinated; for significant periods of time during the *gabinetes* era, Valencia—and not Bilbao or Barcelona—would have likely seemed the only real threat to the hegemony of Madrid.

For example, for their launch in October 1899 Puerto y Novella chose a rather innovative strategy to attract attention to themselves: they challenged Hugens y Acosta to a competition in which cylinders by both *gabinetes* would be played back to an expert panel and then assessed with the aim of declaring a winner. Hugens y Acosta, though, subsequently argued that the event should be held in a theater and the cylinders should be evaluated by the general public and not by experts. Puerto y Novella rejected the offer, as they said that a non-specialist audience would lack the expertise and could be easily misled, and further argued that Hugens y Acosta's refusal of being assessed by experts really hid an awareness that their own cylinders were inferior.[78] In another article, Puerto y Novella also claimed that customers could listen to recordings for free at their *gabinete*—unlike in Madrid, where this service was allegedly offered for a fee.[79] In 1901, *Boletín fonográfico* argued that Valencia was the main capital of the recording industry in Spain, followed by Barcelona and then Bilbao. It certainly might strike us as odd that Bilbao, with a mere two *gabinetes*, was put ahead of Madrid: the writer indeed commented that the situation in Madrid was chaotic, with *gabinetes* competing among themselves and selling at very low prices to the detriment of quality.[80] Again, such claims should not necessarily be taken at face value; indeed, the fact that Madrid's *gabinetes* seemingly responded to each other's strategies swiftly can be regarded as a sign that the market was healthy, rather than the other way round. Still, these sources confirm the regional divides that existed in the *gabinetes'* business.

Gramophone and the end of the *gabinetes*

Further support to the notion that the Valencia industry was not as established as *Boletín fonográfico* often suggested can be gleaned from the fact that both the magazine and the *gabinetes* seemed to go out of business rather abruptly: overall, the early Valencian recorded industry was active for less than two years. Pallás closed his *gabinete* in March 1901. He announced at that point that he intended to stay in the industry, but there is no evidence that he did.[81] *Boletín fonográfico* itself ceased publication rather abruptly too: throughout the first half of 1901, articles and news about photography gradually replaced those on recordings and recording technologies (because, allegedly, many phonograph aficionados were keen photographers too, and vice-versa). During the summer, only a handful of issues were published and, after one single issue in October, the magazine was not published anymore, with no announcement being made in the magazine itself or in the local press about why that was the case. Notices about the closure of *gabinetes* were not published in the press either, but evidence suggests that they were not open long beyond that as such, with the latest notices about them dating from late 1901.[82] Around that time, *El cardo* claimed that, while the industry had picked up in Madrid after the customary deceleration during the summer, the situation in Valencia seemed to have stagnated.[83] In local guides published in 1903 and 1904, Hijos de Blas Cuesta and Vicente Hércules advertised simply as drugstores, with no mention to phonographs or recordings.[84]

Hijos de Blas Cuesta, however, was back in the recording industry only a few months after that—this time as a reseller of Gramophone discs. An article in *Las Provincias* announced the Cuestas' new business venture with a distinct tone of technological inevitability: the journalist wrote that, when the Cuestas were presented with a gramophone by a traveling salesman, they were reluctant at first, because their previous experience suggested that the sounds produced by the gramophone "were like catfights, crickets rather than human voices, and they could destroy the least delicate ears."[85] This time, however, it was the sound of Caruso's recordings of *Mefistofele* that fully persuaded them to become Gramophone representatives.[86]

The Cuestas' story, however localized, is reflective of developments worldwide, with the advent of celebrity recordings gaining new adepts for recording technologies. The concluding comment by the journalist ("The advances in the talking machine are a giant step toward truth; it is, one can say, the truth itself")[87] is similarly typical of narratives of technological inevitability surrounding the phonograph and the gramophone, with the latter presented as inevitably replacing the former on account of its superior capabilities despite initial resistance: indeed, *Boletín fonográfico* had not been as vocal in its defense of the phonograph versus the gramophone as *El cardo* was, but it certainly published some dismissive comments about the gramophone's alleged poor sound quality in its early years.[88] Other *gabinetes* formerly active in the provinces also became Gramophone representatives around this time: Emilio Zurita in Gijón,[89] Prudencio Santos Benito in Salamanca,[90] Francisco Rigau in Tarragona,[91] Casa Erviti in San Sebastián, and Enrique García in Bilbao.[92] The extent to which these individuals and companies managed to shape and disseminate discourses around recorded music was likely considerably more limited than what the Valencia and Madrid *gabinetes* managed to do. Still, we can presume that the expertise and customer networks they had amassed in the *gabinetes* era informed in important ways their activities as multinational resellers—if anything, because it was them who introduced in their locales for the first time the transformative notion that recorded sound was not merely a scientific curiosity, but an aesthetic commodity too.

6

(Dis)embodied voices

Recording singers, 1896–1914

The trope of the disembodied voice is one of the most widely cited in the early history of recording technologies. Emerging from the theorization of sound reproduction as a historical subspecies of writing,[1] this trope was indeed central in Adorno's influential writings on recording technologies[2] and ubiquitous especially in English-language literature,[3] evoking tropes of dehumanization connected to either mechanization and mass production, or to the eerie and supernatural. As I have discussed in Chapters 1 and 2, this trope was absent from Spanish literature and philosophy of the time (with Spaniards being instead more concerned with the phonograph's abilities to tell the truth), but the *gabinetes*' efforts to make sure that their labor was audible in their recordings inadvertently led to a different kind of disembodiment here. Indeed, in the surviving cylinders we can hear the literal voices of some singers from the time, but, even when taking into account written and visual sources of the period, we can hardly hear their metaphorical voices: compared to *gabinete* owners and even to some recording collectors (who will be discussed in Chapter 7), we know considerably less about how performers reacted to the arrival of recording technologies, what their motivations were for recording or for refraining from doing so, what they thought about their own sound when recorded, or what opinions they held about the *gabinetes* industry more generally. First-hand testimonies are practically non-existent, and, even though *Boletín fonográfico* ran features about singers in every issue, providing details about their careers and the *gabinetes* they recorded for, the magazine very much represented the interests of *gabinetes* and not those of the performers they hired, so they were unlikely to feature any critical opinions from the latter.

By examining what we know about the *gabinetes*' hiring and recording routines and pricing practices, the singers' biographies and careers, and the sounds we can still hear in some of the cylinders, this chapter discusses the role of singers in shaping the nascent recording industry in Spain and, in turn, how the newly developing recording industry influenced their careers and on the music profession. Although I will engage in discussion of the musical features of a few selected cylinders, this chapter is not mainly concerned with performance practice and instead inserts itself within the cultural history approach which is dominant in

Inventing the Recording. Eva Moreda Rodríguez, Oxford University Press. © Oxford University Press 2021.
DOI: 10.1093/oso/9780197552063.003.0007

this book. Discussion of the music is therefore oriented principally toward further comprehending the singers' role in the nascent recording industry; similarly, further discussion of music and performance will take place in Chapter 7, this time oriented toward ascertaining the particularities and the dynamics of home recording in this era. At the same time, I hope that a more in-depth understanding of the singers' place and role in the nascent industry, as I try to present in this chapter, will encourage other researchers to embark in more thorough explorations of performance practices as documented in those cylinders in the future; this is not, however, the main aim of this chapter or this book.

In the previous paragraph, I have deliberately used the word "singers" as opposed to "performers" or "musicians," since I believe that the previously cited questions are most urgent when applied to recordings of vocal music and, particularly, to the musical-theatrical genres of opera and *zarzuela*. There are three reasons for that. The first and most pragmatic is that a majority of the surviving *gabinetes'* recordings are indeed of vocal music (mostly opera and *zarzuela*), whereas just under one third are instrumental (mostly wind bands). Moreover, *gabinete* owners indeed dedicated more efforts to vocal recordings than to instrumental ones. Although *gabinetes* did not always employ well-known or competent singers, many of them at least attempted to do so, whereas instrumental recordings do not seem to have attracted the same level of attention or care: for example, *gabinetes* did not normally hire concert pianists to record solo pieces but instead it tended to be the *gabinete's* staff accompanist who would record most of the piano cylinders on offer in the catalogue, from selected examples of classical pianistic literature (mainly Chopin) to salon music (polkas, marches, *pasodobles*). This was the case with José Bellver for Hijos de Blas Cuesta, señor Armengol for Fono-Reyna, and señor Pérez Soriano for Viuda de Aramburo; there is no evidence that any of them was particularly well known or distinguished as a concert pianist in those days. Only Hugens y Acosta recorded some relatively more prestigious instrumentalists, such as chamber ensembles drawn from the Sociedad de Conciertos in Madrid—the first regular symphony orchestra in Spain. Singers were also recorded to higher standards than instrumentalists: while it is true that wax cylinders were at the time more apt to record the former rather than the latter, *Boletín fonográfico* also suggests that *gabinetes* owners and phonograph buyers were mostly concerned with developing new techniques to record voices, but instruments were lower in their list of priorities. In the second-ever issue of the magazine, a Mr. Marín published an article with advice for making home recordings[4]—but this focused exclusively on the voice, identifying what he thought should be the ideal recorded sound:

> She had impressed all the notes with remarkable sweetness, with a big sound, with no stridencies or shrillness; both the high and the low notes were heard

with such clarity that it would seem the instrument was next to me. (The voice) was limpid and pure, and there was no stridency or artificial darkness of any type, either in the low, middle or high notes.[5]

Marín also gave detailed advice on how to achieve such standards, including the positioning of the phonograph and the singers, and the types of furniture and rugs that should be avoided.

A second reason why it makes sense to focus on vocal recordings is the cultural significance that vocal genres had in Spanish musical life at the time, particularly among the bourgeoisie able to acquire phonographs and cylinders. As I discussed in Chapter 2, what drew Spaniards to recording technologies in the era of the traveling phonograph was the possibility of listening to sounds which were already part of their everyday soundscape, and the *gabinetes* did little to change that. It is therefore no surprise that about a quarter of the surviving cylinders are of opera: bel canto, French *grand opéra* (typically recorded in Italian, as it was sung in the opera theaters of the time), Verdi and some of the latest successes of *verismo*, and, to a lesser extent, Wagner. Moreover, with debates proliferating in the second half of the nineteenth century among composers and audiences about the significance and development of Spanish-language opera, it is also no surprise that Emilio Arrieta's *Marina* (1855) and Tomás Bretón's *La Dolores* (1895) were—if we go by the surviving cylinders— among the most frequently recorded operas too. After opera there followed *género chico*, which by 1900 had been firmly established as *zarzuela*'s most popular sub-genre for about twenty years. It consisted of one-hour plays typically on a light or comic subject, with Madrid theaters putting on four different plays every evening (the so-called *teatro por horas*, or theater by the hour model). The *gabinetes*' catalogue reflects to a considerable extent the rapidly changing nature of the genre: about half of the surviving recordings are of works premiered during the same time period when the *gabinetes* were active (1896–1905), and the works for which more recordings have survived largely coincide with those which had greater success on stage: *La viejecita, La revoltosa, El tambor de granaderos, Gigantes y cabezudos*, together with a few works that preceded the *gabinetes* era but were still regularly performed and increasingly regarded as classics of the genre (*El santo de la Isidra, El dúo de la Africana*). Flamenco and *zarzuela grande* (the full-length, serious form of *zarzuela*, which was at its peak in the years 1850–1880 and whose repertoire works were still regularly performed around 1900) were also relatively well represented, with some 90 and 70 surviving cylinders respectively. A minority of cylinders cover other vocal genres, such as traditional music from other parts of Spain (Aragonese *jotas*, Basque *zortzikos*) and Spanish-language art song (not French or German, which were still rarely heard in Spain).

A third reason for focusing on vocal recordings lies in the close historical link that tied to the phonograph and the human voice from the moment the first was invented. According to Lisa Gitelman, the phonograph originated as a *language machine*—in a context in which interest in language, under the form of phonology and phonetics, was increasing.[6] The intended uses of the phonograph listed by Edison in 1877 did not focus on music, but they were all concerned with human speech (mostly under the form of dictation and the preservation of the human voice), and this focus on the human voice remained when he launched the Perfected Phonograph in 1888.[7] In speculative commentary throughout the 1880s and early 1890s about the potential uses the phonograph, Spaniards mostly followed Edison's assumption that the phonograph's potential lay in its ability to record the human voice in its different forms for a multitude of practical purposes[8] (even though, contrary to other countries,[9] none of these potential practical uses—dictation, teaching, postal communication—really caught on in Spain), and in Chapter 2 I examined how most recordings played back by traveling phonograph demonstrators indeed featured the human voice. The *gabinetes* introduced a new notion: the idea the phonograph's main potential resided in entertainment and not administration. This still predominantly involved the human voice rather than instruments or other types of sounds. Several *gabinetes* released "cuentos" (short stories, more akin to jokes and anecdotes) and satirical speeches, but no serious speeches or theatrical passages, even though such uses had been suggested in the early years after the invention of the phonograph. Spoken language recordings, however, are only a small fraction of surviving cylinders and they featured much less prominently than music ones in the *gabinetes'* publicity and catalogues, suggesting that they were less popular and eventually disappeared. By the time Gramophone settled in Spain, its strategy focused on the device's ability to capture the human singing voice, and spoken word recordings hardly featured.[10]

Recordings and the common singer

In Chapter 2, I discussed how some of the *salones fonográficos* proliferating in Spain during the 1890s engaged well-known singers such as *género chico tiple* Lucrecia Arana, and in Chapter 3 I also explained how some of the *gabinetes* hired prominent opera singers (rather than *género chico* ones, who would have been seen as less prestigious) for their launch parties, during which their voices were recorded and played back to the audience. Nevertheless, once *gabinetes* started recording commercially, their catalogues tended to rely on lesser-known singers rather than celebrities. There were, of course, exceptions, some of which will be discussed in the course of this chapter, but this mirrors trends elsewhere in the world, with working but not excessively popular singers making up the lion's share of catalogues before "celebrity recordings" became popular in 1903–1904, following Enrico Caruso's example.[11] We can imagine that the reasons that drove well-known Spanish singers away from *gabinetes* were roughly comparable to those elsewhere. These had mostly to do with the labor and complications involved with recording cylinders one by one. As is well known from other accounts of the earlier days of recording technologies, singers often had to walk from one spot to another and even contort during the recording process so as to optimize the sound signal that was captured by the phonograph,[12] which many might have regarded as inconvenient and unbecoming. Singers might have also been worried that the phonograph did not capture their voices to the highest level of fidelity, and they might have harbored concerns about the recording industry, which, as a new developing sector, was not fully established in terms of reputation and reliability. Many well-known singers might have therefore initially regarded recording processes as cumbersome and unrewarding, both financially and artistically, particularly if compared with stage engagements. As the quality of recordings improved and the industry became more established, particularly with the arrival of Gramophone, well-known singers became more interested in recording and opportunities for lesser-known ones decreased. We can presume that similar processes took place worldwide, although with local differences. Singers with local careers were still prominently represented in Edison catalogues of the mid-1900s,[13] with recordings therefore presumably acting as a launchpad for such singers to reach the international market; however, in Spain such opportunities disappeared a few years earlier as the *gabinetes* closed in 1903–1905.

There is evidence indeed that the *gabinetes* might have struggled to attract prominent singers. *Género chico*'s *primeras tiples* (equivalent to opera *prima donnas*, with a stable contract at one of the main theaters and commanding high salaries), for example, were not particularly well represented among their catalogues. There is no evidence that Luisa Campos, Isabel Bru, Joaquina Pino, and Irene Alba recorded for the *gabinetes*. On the other hand, Lucrecia Arana, Matilde Pretel, Leocadia Alba, and Concha Segura all did. Unlike some lesser-known singers who recorded for a multiplicity of *gabinetes*, though, they only recorded for Hugens y Acosta,[14] which, being the best-established label at the time, could presumably pay higher prices and offer enhanced technical capabilities. In any case, very few *gabinetes* recordings of these four singers have survived (four for Alba, two for Pretel, one for Arana, none for Segura), which suggests that they might have recorded less often than their lesser-known counterparts. Arana, however, did record repeatedly for Gramophone from 1904 onward, suggesting that she did overcome her reservations once the industry became better established both technologically and commercially.

This does not mean, however, that either *gabinetes* or their customers regarded all singers as equal regardless of whether they were well known or not. A singer's fame and level of recognition would typically be reflected in the prices of his or her recordings: whereas famous operatic baritone and singing teacher Napoleón Verger, recording for Hugens y Acosta, commanded 50 *pesetas* per cylinder,[15] the *género chico* cylinders made by Amparo Cardenal—an amateur teenager who developed a modest recording career work for Puerto y Novella in Valencia—were sold for no more than four. However, prices did not always reflect in a straightforward manner the prestige that a singer had previously acquired on stage, as illustrated by tenor Julián Biel. Biel, born in Zaragoza in 1870 and originally a member of the Teatro Real choir, recorded his first cylinders for Hugens y Acosta after failing to launch a solo stage career.[16] It was at the Hugens y Acosta premises that Biel met Maestro Almiñana, who hired him for the summer opera season of the Teatro del Buen Retiro. In the summer of 1899 Biel sang main roles in *Hernani*, *Cavalleria rusticana*, *Pagliacci*, *Il trovatore*, and *L'africaine* there.[17] In the Hugens y Acosta catalogue for 1900, Biel's cylinders were on sale for 30 *pesetas*[18]—the most expensive for operatic tenors, and at higher prices than singers who had been established in the Madrid theaters for much longer, suggesting that the *gabinetes'* pricing strategy could have both reflected and influenced temporary fashions for one singer or another. Further factors influencing the pricing policy of the *gabinetes* included genre (opera tended to be the most expensive genre, followed by *zarzuela*, whereas flamenco and traditional music cylinders could be sold for as little as 1.5 or 2 *pesetas*) and voice type (with tenors generally commanding the highest prices, followed by sopranos, and contraltos and basses commanding the lowest).

In the same way as it is difficult to ascertain the singers' motivations to record for the *gabinetes* or refraining from doing so, there are not many sources either which allow us to reconstruct how the day-to-day business of recording would look from the point of view of a singer. Biel is indeed one of the few singers for whom we have an indication about how much he earned per session (4 *pesetas* per cylinder), as well as how often he recorded (every three days).[19] This was, however, before he achieved success as a soloist on the stage, after which we can presume that his wages increased and the frequency of his visits to Hugens y Acosta would have decreased as his stage commitments became more numerous and, presumably, better paid. In his 1900 catalogue, Hugens y Acosta claimed that it was not possible to secure top *zarzuela* singers for the recording studio for 2 *pesetas* per cylinder.[20] This claim makes sense if considered in the context of singers' wages at the time. At the time, a *zarzuela primera tiple* normally earned from 50 to 100 *pesetas* per performance.[21] While rehearsal time was unpaid and singers had to provide their own costumes, the "economies of scale" model that dominated *género chico* meant that the industry could be reasonably profitable for a singer as long as the works they took part in stayed on the stages for at least a few weeks. Of course, while recording might not have been a financially appealing proposition for *primeras tiples*, less established singers with inferior salaries, or those who had not been in a successful production for a while, might have been tempted to accept lower salaries than the 2 *pesetas* suggested by Hugens y Acosta.

Boletín fonográfico and the general press occasionally published notices about singers who claimed to be able to make a living solely off recordings. This is the case with former *género chico* tenor Jesús Valiente, who claimed in 1900 that he had recorded 7,000 cylinders for Puerto y Novella in 1900, and that this had allowed him to leave stage engagements behind.[22] Such statements, however, must be examined cautiously: Valiente was self-admittedly an exception in the *gabinetes* era, and his foray into being a full-time recording performer did not last long anyway: he was back on the stage in Santander in 1903,[23] after Puerto y Novella (and most other *gabinetes*) ceased trading. More generally, with recordings supplementing and building upon the live theatrical experience (as has been discussed in previous chapters), evidence suggests that, for singers, visits to the *gabinetes'* recording studios could indeed provide some extra income and visibility, but would not replace success on stage. It is telling in this regard that the magazines which occupied themselves with *zarzuela* at this stage, informing readers about the latest and coming premieres and including a fair amount of gossip about singers, composers, and impresarios (*Juan Rana*, *El arte de el teatro*) did not pay any attention to *gabinetes'* recordings or to the phonograph as a phenomenon, suggesting that recordings were seen at most as an extra or a curiosity, but not a serious competitor to the live theatrical experience. In other words, a singer still needed to prove himself or herself on the stage if he or she desired to have a career.

Even though first-hand testimonies from singers about the recording industry are extremely rare, delving into aspects of their working lives at this time can indeed provide some insights into how they might have regarded the nascent recording industry.[24] Well before the advent of the phonograph, singers were well used to juggling a variety of engagements and sources of income in their attempts at getting ahead in what could be a fluctuating, difficult market. We certainly know, for example, that the working lives of many singers were characterized by mobility: opera and *zarzuela* theaters were typically rented by the season, with the theater's board of directors issuing a call for expressions of interest during the summer and companies sending in bids; the winning company would earn the right to run the theater during the coming season—until the following summer, when the process would start again. Some of the most reputed companies rotated between the various theaters in Madrid; lesser prestigious companies were to be found in the provinces, occasionally securing engagements in Madrid and Barcelona. Smaller towns did not have a regular company throughout the years and were staffed instead by touring companies with no secure "home" for the year.

In the same way as companies moved between theaters, performers moved between companies too—either by necessity or by choice. Within a company, singers were typically hired to fulfill particular kinds of roles according to their singing and acting skills and capabilities; a *primera tiple* in a *género chico* company would be expected to have a modicum of voice, as well as strong acting skills and a pleasant stage presence so that she could be convincing in leading roles, whereas a *tiple característica* would not always be expected to have a beautiful and trained voice, but she would be expected to sing expressively and reliably providing comic relief throughout a performance. Singers did not always stick to the same types of roles throughout their careers, moving up and down as their skills developed or changed. Sometimes, they moved between genres too, including opera, *zarzuela grande, género chico, género ínfimo,* and *cuplé,*[25] and spoken theater. In this context, while a few well-established figures, such as Lucrecia Arana, could command high salaries, intervene in the decisions of the companies they sang for, and choose which roles and types of music to perform, the overwhelming majority of *zarzuela* singers were considerably less fortunate. In this context, many singers might have regarded recordings as a welcome additional source of income and visibility that they could juggle with a variety of stage engagements.

Since *zarzuela* and opera in Spain have not yet been studied from a music profession point of view, we have yet to fully understand the sorts of decisions and strategies performers engaged in, as well as the pressures put on them, to move between different roles, responsibilities, and genres, and, consequently, to comprehend how making recordings would have fitted among these options. But there are some insights that we can gain from looking at some of these individuals' trajectories and examining whether they made recordings, what they recorded, and at which point in their careers they did so. Leocadia Alba provides an illustrative example—although we must be careful too not to generalize from her example, since long and successful careers such as Alba's were not the norm. Born in 1866 in a theatrical family, Alba first appeared on stage in her teens, and, after a series of minor roles in touring companies, she established herself as an in-demand *primera tiple* in *género chico*. She performed Angelita in the 1887 premiere of Manuel Fernández Caballero's *Chateau Margaux*[26]—a relatively demanding role, requiring a reliable top register as well as some coloratura abilities. Alba's surviving cylinders are considerably deteriorated, but even through the noise it is possible to ascertain that Alba's diction was clear and her voice appears to be that of solid singer with a lyric coloratura voice and vibrant high notes.[27] However, at the time Alba recorded these cylinders she had already moved in a different direction, undertaking mostly *tiple característica* roles. The reason for this career change was that she allegedly saw more of a niche here than in the competitive market for *primeras tiples*; moreover, with *género ínfimo* starting to develop, such *primeras tiples* were increasingly required to wear revealing clothes on stage, which Alba felt did not suit her physique.[28] The fact that the surviving recordings of Alba's are in the *primera tiple* rather than *tiple característica* category certainly pose questions surrounding the introduction of recordings into the music profession and their impact into individual careers. A plausible scenario here is that *gabinetes* preferred to record *primera tiple* rather than *tiple característica* numbers as the latter were scarcer and less interesting vocally: they capitalized instead on the performer's text delivery, acting, and movement skills, and often on her physique too, so recording the music at the exclusion of everything else might have considerably diminished their appeal. At the same time, though, Alba still had the vocal skills necessary to sing this repertoire, and she might have appreciated being given the opportunity to do so even though she had abandoned such roles on stage. Alba eventually finished her career as *actriz de verso*, that is, as a spoken theater actress.

Questions surrounding the day-to-day mechanics of recording and the relationship between the singers and the *gabinetes* are, again, difficult to answer conclusively, but existing recordings and written sources provide some useful directions. It can be presumed that, at least in Madrid, initial contacts and conversations could have happened in the theaters themselves, given that *gabinetes* tended to be in the vicinity of those. This, which was discussed as a strategy to attract customers in Chapter 3, could work equally well in terms of having easy access to singers.[29] Some evidence for this comes from the numbers of singers employed by each of the *gabinetes*. Whereas some *gabinetes* in Madrid—particularly Hugens y Acosta and, to a lesser extent, Fono-Reyna and Viuda de Aramburo—availed themselves of a multiplicity of performers, those in Valencia (Puerto y Novella, Hijos de Blas Cuesta) seem to have operated a more structured system, similar in some respects to theatrical companies, with one singer per voice type and genre. Madrid, of course, was home to a multiplicity of *género chico*, *zarzuela grande*, and opera theaters with permanent companies, as well as aspiring singers, whereas Valencia *gabinetes* had to rely mostly on touring companies. Puerto y Novella, for example, systematically employed one operatic soprano (Josefina Huguet), one operatic mezzo (Inés Salvador), one *zarzuela tiple* (Amparo Cardenal), and three tenors: one for *género chico* (Jesús Valiente), one for *zarzuela grande* (Manuel Figuerola), and one for opera (Gabriel Hernández).

Conversely, singers in Madrid had access to a greater range of *gabinetes* they could offer their services too, in search for better conditions and wages. However, not many seem to have taken advantage of such possibilities: most performers are only known to have recorded for one or two *gabinetes*, and only a minority of them were considerably more active. These include flamenco singer El Mochuelo, who recorded for at least six *gabinetes* (four in Madrid and two in Barcelona), *zarzuela* baritone Emilio Cabello, *tiple* Eloísa López-Marán, operatic tenors Francisco G. Pertierra and Rafael Bezares (four *gabinetes* each), and operatic soprano Josefina Huguet (three *gabinetes* in Barcelona, Madrid, and Valencia). While Huguet was a well-established international opera star at the time and Bezares was developing an appreciable stage career in *zarzuela grande* and opera, others had more modest stage careers—including El Mochuelo, who only really came to prominence in Spain because of his prolific activity making recordings. While we might presume that the reasons who took these specific singers to be especially active in the recording arena were personal and unique to a considerable extent, what we might hypothesize is that they all had a pioneering interest in recordings, and/or they acquired a familiarity with and fondness of the earlier recording processes which made a more attractive proposition to *gabinetes* than their less experienced colleagues.

The list of singers who recorded for more than one *gabinete* overlaps to a significant extent with the list of those of whom the most recordings have survived, as suggested by Table 6.1. The picture these details paint is one where a relatively small number of keen singers constituted the core of the Spanish recording industry, with many others taking up more occasional engagements. In the remainder of this section, I will therefore look in detail at the profiles at some of the most recorded singers, covering the full range from highly successful ones to amateurs, in an attempt to draw further conclusions that can guide our understanding of these recordings mainly as sources for the history of recording technologies in the Spanish context. I will be also making some observations regarding performance practices, mainly when this is relevant to understand the broader context in which singers worked, even though, as stated at the beginning of this chapter, it is not the aim of this book to provide detailed discussion of performance practice.

Operatic and *zarzuela grande* tenor Rafael Bezares is, without a doubt, the singer of whom the most cylinders have survived: no fewer than 36, including Spanish opera (the usual *La Dolores* and *Marina*), Spanish-language song ("La partida" and "Canto del presidiario," both popular works by Fermín María Álvarez), and some *zarzuela grande*. Twenty-seven of the Bezares cylinders are held at the Ybarra family collection held at Eresbil Archivo de la Música Vasca.[30] As I will discuss in Chapter 7 more extensively, the make-up of the private collections that have survived can sometimes skew our conclusions about who was recording in Spain and what they were recording around this time: the fact that Bezares is the singer of whom the most recordings have survived might simply indicate that he was a favorite of the Ybarras, or that he was indeed among

Table 6.1 Performers (*zarzuela* and opera) with the highest number of surviving commercial recordings

Performer	Specialization	Number of recordings[a]
Rafael Bezares	Opera tenor	36
Señor Navarro[b]	*Género chico* comic tenor	22
Jesús Valiente	*Género chico* tenor	18
Josefina Huguet	Opera soprano	17
Amparo Cardenal	*Género chico* tiple	16
Inés Salvador	Opera mezzo	16
Señor Romero	Opera and *zarzuela grande* baritone	13
Emilio Cabello	*Género chico* tenor	12
María Galvany	Opera soprano	12
Lamberto Alonso	Opera tenor	11
Eloísa López-Marán	*Género chico* tiple	11
Luisa García Rubio	Opera soprano	8
Ignacio Varela	Opera tenor	8
Julián Biel	Opera tenor	7
Blanca del Carmen	*Género chico* tiple	7
Señor Velo	Opera tenor	7
Ramón Blanchart	Opera baritone	6
Francisco G. Pertierra	Opera tenor	6
Marino Aineto	Opera baritone	5
Bernardino Blanquer	Opera and *zarzuela grande* tenor	5
Ángel Constantí	Opera tenor	5
Manuel Figuerola	Opera and *zarzuela grande* tenor	5
Ramona Galán	Opera mezzo	5
Señorita Martínez	*Género chico* tiple	5
Francisco Souza	Opera baritone	5
Angelo Spagliardi	Opera tenor	5
Antonio Vidal	Opera bass	5
Leocadia Alba	*Género chico* tiple	4
Manuela Cubas	*Género chico* tiple	4

Table 6.1 *Continued*

Performer	Specialization	Number of recordings[a]
Gabriel Hernández	Opera tenor	4
Señor León	Opera bass	4
Anita Lopeteghi	Opera soprano	4
Señorita Rossi	Opera soprano	4
María Vendrell	Opera soprano	4
Antonio Domingo	*Género chico* tenor	3
Lolita Escalona	*Género chico* tiple	3
Luis Iribarne	Opera baritone	3
Juan Romeu	*Zarzuela grande* and opera baritone	3
José Sigler	*Género chico* baritone	3

[a] Includes both solo and ensemble recordings (duet, tercet, etc.), so some recordings might have been counted twice.

[b] Navarro is a relatively common surname, and at least three singers were active in these years that could have plausibly made the recording: Enrique Navarro, José Navarro (both comic tenors), and Luis Navarro, whose vocal range is not known (María Luz González Peña, "Enrique Navarro," "Luis Navarro," and "José Navarro," *Diccionario de la zarzuela. España e Hispanoamérica*, ed. Emilio Casares (Madrid: ICCMU, 2008), 1:339.

the most prolific of his time. Other evidence suggests the latter is true: Bezares was prominently featured in catalogues of the time, and it is also significant that Bezares was mentioned in the press at least twice in connection with recordings. Indeed, as anticipated in Chapters 3 and 5, critical commentary on the musical and artistic value of recordings was rare at this time: articles in *Boletín fonográfico* and *El cardo*, as well as the general press, focused on the technological qualities of cylinders, but rarely discussed the repertoire, the singer's abilities, or her suitability to be recorded. Very few exceptions exist, and the fact that two of them mentioned Bezares suggests that the tenor was indeed starting to make a reputation for himself as a recording artist. The first example was published in 1901 under the pseudonym Christián de Neuvillette. The article, a short feature about Bezares, names his recordings at a time where coverage of singers in the general press did not normally do so. More tellingly, the writer also asked: "Who hasn't heard one of the cylinders that Bezares has impressed for one or another *gabinete fonográfico*? His voice is small, without volume, sometimes veiled, but always sweet and well managed."[31] The second example was published in 1905 in *El diario de Córdoba*. It is a poem allegedly written by a group of Bezares fans who admitted they mostly knew his voice from the recordings he had made.[32] Such admissions were, again, rare at the time.

Although Bezares did appear in the Teatro Real alongside María Barrientos in 1901 as Elvino, his career also included stints in less prestigious theaters, and in 1905 he switched from opera to *zarzuela grande*.[33] Many of the surviving recordings, particularly those at the Ybarra collection, have deteriorated over time and might not offer an accurate representation of his abilities. They suggest, however, that Bezares's middle register was weak (as was sometimes the case with *zarzuela grande* tenors); his high register, however, was vibrant, bright, and somewhat nasal, and would have likely been seen as recording well. Bezares also followed the performative conventions of his time in making moderate use of portamentos and fermatas to convey expressiveness.[34]

With 17 surviving commercial cylinders, Catalan soprano Josefina Huguet was the most prolific opera singer after Bezares; she also recorded 23 cylinders privately for Catalan textile industrialist Ruperto Regordosa, who will be duly discussed in Chapter 7. One reason why Huguet recorded so prolifically might have to do with her voice type. Huguet was a light-lyric coloratura soprano with a very clear and pure high register, and such voices recorded particularly well on wax cylinders,[35] as is obvious from the surviving recordings: among the female singers with the most surviving cylinders is also María Galvany, who had a similar voice type.[36] By the time the *gabinetes* started to record, Barcelona-born Huguet had a decade-long career in prestigious opera theaters behind her; this no doubt contributed to the prices of her cylinders being among the most expensive for operatic sopranos, reaching 25 *pesetas* at Puerto y Novella.

Huguet's recordings provide valuable insights in three respects: firstly, on how recordings might have influenced the careers of singers; secondly, on what sorts of meanings and functions listeners might have attributed to recordings at this time; thirdly, on how recordings might have influenced the repertoires that singers performed and recorded. (They, on the other hand, can also provide useful insights about performance practice, which is not the main topic of this chapter: the three themes outlined previously, however, provide a foundation to understand the significance of these cylinders as documents of performance practice.) In which concerns the influence that recordings exerted on the careers of singers, from the available evidence Huguet emerges as one of the first performers, in Spain and worldwide, to have had a sustained recording career which stretched well beyond the *gabinetes* era. Indeed, when the Compagnie Française du Gramophone started regularly visiting Madrid and Barcelona to record local artists from 1899 onward, Huguet was among the few Spanish opera singers to be recorded (another was Bezares). In Spain, the company focused on indigenous genres (*zarzuela*, flamenco), which was in line with the new industry trend of recording local or "ethnic" repertoires to both cater for strongly localized markets while sending some exoticism elsewhere.[37] Gramophone preferred to record opera singers in well-known operatic centers such as Milan and Paris, so the fact that Huguet was selected to record repertoire in Barcelona which could have easily been recorded elsewhere suggests that the company saw her as a prestigious singer with good recording abilities—not as one more Spanish soprano who should limit herself to her own native repertoire. Years after the *gabinetes* era, Huguet's recording of *Lucia di Lammermoor*'s "Regnava nel silenzio" was released in the prestigious Victor Black Label series in 1907 (under the name Giuseppina Huguet: it was common for Spanish artists in those days to switch between Spanish and Italian first names),[38] with other recordings were released in the less prestigious Victor Blue Label.[39] Huguet's trajectory suggests that she was one of a minority of singers with successful stage careers who nevertheless developed an early interest in recordings, as well as an ability to record well; as such, they were in demand from both *gabinetes* and multinational companies. Unlike with lesser-known singers, Huguet managed to remain in demand once multinationals centralized the recording market, because she had worldwide prestige thanks to her stage career. Singers with comparable trajectories to Huguet include those who recorded for the prestigious Milan-based label Fonotipia (baritone Ramón Blanchart, who recorded for Hugens y Acosta and also for Edison, and bass Andrés Perelló de Segurola, who recorded for Álvaro Ureña and Gramophone) and for Odeón,[40] (operatic tenor Florencio Constantino, soprano Anita Lopeteghi, and *zarzuela* performers Manuel Figuerola and señorita Santisteven).[41] Huguet also recorded privately for Ruperto Regordosa in Barcelona, and these cylinders reveal further details of her approach to the recording process, as will be discussed in Chapter 7: certainly, Huguet is perhaps the only Spanish singer we can follow relatively closely during the early years of the industry.

Huguet's recordings also help us understand the changing and evolving attitudes of listeners toward recordings at this time. Indeed, even though Huguet recorded rather extensively, she did so in a rather small repertoire, and recorded the same arias several times. Some of these recordings show considerable variability with respect to each other: for example, Huguet systematically introduced cadenzas in her recordings, and she changed them from one recording to another, while at the same time keeping some core structural elements.[42] This suggests that recordings at the time did not necessarily strive to be perfect versions of any given piece or aria, but rather tried to capture some of the variability inherent in live performance.

The third area of interest in Huguet's recordings concerns how these might have influenced and shaped the repertoire she chose to sing. While the core of Huguet's repertoire was in her light-lyric coloratura roles, she also recorded arias from some roles that she is not known to have sung on stage, such as both of

Mimi's arias from *La Bohème and* excerpts from *Lohengrin*. These were both among the most-recorded operas of the *gabinetes* era, so it is possible that Huguet was asked to record those for commercial rather than artistic reasons. We should therefore ask ourselves how valuable these recordings might be as documents of performance practice, when it is unlikely that the singer ever performed these arias on stage.

Huguet was certainly not the only singer to record outside her specialization. The aria "Quando m'en vo," again from *La Bohème*, provides perhaps the most illustrative example: it was among the most recorded in the *gabinetes* era, and different kinds of singers sung it, ranging from mezzo Inés Salvador for Puerto y Novella (transported down by one tone),[43] to light-lyric coloraturas such as Huguet and María Vendrell, recording for Hijos de Blas Cuesta.[44] Men occasionally recorded outside their Fach too, as is the case with tenor Ángel Constantí, who sang mostly lyric and spinto operatic roles on stage.[45] Constantí did record the repertoire he was better known for both commercially at Barcelona *gabinetes* and privately for Regordosa, but he also recorded Wagner in both contexts, which there is no evidence he ever sang on stage.[46] Throughout his recordings, Constantí certainly sounds more at ease in his usual stage repertoire, in which he introduces subtle tempo changes for expressivity, in line with recording practices of the time; these partly make up for the lack of dynamic range that could be captured in cylinders.[47] Constantí's Wagnerian recordings can introduce some nuance in our understanding of how *gabinetes* might have selected repertoires and singers. Indeed, as I suggested in Chapter 4, it is surprising that, with the Liceu being a well-known Wagnerian hub, only these two commercial recordings of Wagner by Barcelona *gabinetes* have survived. It is plausible that both Manuel Moreno Casas and Sociedad Artístico-Fonográfica sensed in this repertoire a business opportunity that Constantí spotted too.

Whereas Huguet, Salvador, and Constantí only occasionally departed from their Fach's usual repertoire for recording purposes but still stuck mostly to opera, other singers moved more easily between genres when going into the recording studio. This is the case with Lamberto Alonso, Bernardino Blanquer, and Eloísa López-Marán, among others. Alonso, a native of Valencia, did have singing training and had a brief and auspicious theatrical career in Barcelona and Madrid before he decided to return to his hometown and earn a living by being professionally employed at a church there.[48] For his alleged lack of dedication to cultivating his talent and pursuing a leading career, *Boletín fonográfico* mockingly termed Alonso "a criminal of art."[49] Alonso, however, enjoyed a modicum of fame after the advent of the phonograph in Spain and recorded extensively for Hijos de Blas Cuesta. His surviving recordings cover a range of lyric to dramatic tenor repertoire, from *Carmen*, *La favorita*, and *La Bohème* to *La Gioconda* and *Lohegrin*. Although the heavier repertoire did not suit Alonso particularly well and his vibrato was limited, he reveals itself in these recordings as a performer with a bright, rather unusual timbre.[50] It might have been Alonso's relative success with his Blas Cuesta recordings that allowed him to revive his stage career: in 1901 he sang at the premiere of *El fantasma*, by local composer Salvador Giner, in Valencia and was then awarded a scholarship to further his singing studies in Italy. Upon his return to Valencia, he was active in local musical life, and became influential too as a singing teacher.[51]

Although based in Madrid and not in Valencia, the recording career of tenor Bernardino Blanquer shows some similarities to Alonso's. Blanquer's stage career was, by all accounts, quite modest: he sang *comprimario* roles at the Teatro Real in the 1896–1897 season and at the Teatro Lírico (specializing in *zarzuela*) in 1902,[52] and gave a few recitals together with other singers.[53] In 1906 he was working as a church tenor at the Capilla Isidoriana in Madrid.[54] Five recordings by Blanquer have survived, for La fonográfica madrileña and Hugens y Acosta (where his recordings, at 6 *pesetas*, were among the cheapest tenor ones), of *Marina*, *Rigoletto*, *Pagliacci*, *Cavalleria rusticana*, and one *zarzuela grande*, *La tempestad*. In his recording of *Cavalleria rusticana*'s "O Lola" for Hugens y Acosta, Blanquer's voice appears rather thin and possessing limited vibrato, although his phrasing was certainly expressive.[55] His recording of the tenor romanza "Salve, costa de Bretaña," also for Hugens y Acosta, reveals to an even greater extent an intelligent and expressive performer with a rather small voice but very clear and expressive diction in Spanish.[56]

Eloísa López-Marán (often introduced in recordings as "señorita Marán"), on the other hand, had a short career in which she nevertheless managed to cross over from opera to lighter opera and back. She studied with famed baritone and teacher Napoleón Verger[57] and first appeared with at the Teatro del Buen Retiro in Madrid in 1897. When she appeared in the comic opera *Campanone* (more akin to *género chico*) at the Teatro Parish in February of the next year,[58] *La correspondencia de España* called her a "defector" from opera to *zarzuela*,[59] while *La Iberia* wrote that she had a good voice (without giving further details, which was not uncommon in opera reviews of the time focusing on lesser-known singers).[60] She was employed again at the Teatro del Buen Retiro in Madrid in 1899 and 1900[61] and sang a *comprimaria* role at the Teatro Real in 1901,[62] but there are no further traces of her presence on stage. Eleven cylinders by López-Marán have survived, with three of them being duets with a señor Navarro, a señor Pretel, and José Rovira. With the exception of the duet from *Marina* and the "Alborada" from *zarzuela grande El señor Joaquín*, all other surviving recordings are of *género chico*, which López-Marán allegedly never sang on stage. What this could indicate is there was some demand among *gabinetes* for singers who, like Leocadia Alba, could sing *género chico* romanzas with a trained, strong voice: on stage, a *primera tiple* could often make up for a less-than-ideal singing voice with strong acting and comedic skills, but it is more dubious that this would have transferred comfortably into recordings. López-Marán's classical training is evident from the surviving cylinders and, whereas her diction is not as good as would have been expected of a *primera tiple*, her performance was certainly made expressive by the energy and tightness emanating from her voice, which not all *género chico* recordings exhibited.[63]

Recording artists?

Even though examples such as Bezares, Huguet, and the others I have discussed in the preceding pages offer significant insights about how singers might have combined recording sessions with stage engagements, there is still a small but significant number of performers who recorded for *gabinetes* but are not known to have had a stage career. These included Luisa Alarcón, señoritas Asenjo, Areli, and Berruezo, Pepita López, Teodoro Lánderer,[64] Francisco Cervera, Aníbal Chequini,[65] Roberto Jartbó, and Concha Sanz Arnal, among others. Perhaps to a greater extent than with professional singers, the lack of sources makes it difficult to understand why they might have gone into making recordings and why *gabinetes* might have hired them. Some scattered evidence exists. For example, Concha Sanz Arnal, who recorded only for Hijos de Blas Cuesta, came from a respectable, bourgeois family in Valencia, and as such she disliked the idea of using her voice to entertain individuals other than her own relatives and close friends. In an interview with *Boletín fonográfico* she stated that she found recordings to be an acceptable compromise, in that they allowed her to earn money while allowing her to remain respectable.[66] At the time, indeed, it was not uncommon for young women from the bourgeoisie to study singing and entertain family and guests at home; them going into the theatrical profession, however, was not seen as a respectable career path. It is possible that some of the previously mentioned performers, especially female, had similar trajectories to Sanz Arnal's. Luisa Vela provides a similar, interesting example. Vela started off recording dramatic soprano repertoire for *gabinetes* in Valencia as a teenager with no stage experience, for reasons comparable to Sánchez Arnal's.[67] However, unlike Sánchez Arnal, she was eventually persuaded to launch a stage career and went on to become one of the most famous *zarzuela* performers of the early twentieth century together with her husband, Emilio Sagi. Another reason why *gabinetes* might have chosen to record amateur singers is that they might have been ready to work for lower wages than professionals: indeed, several of the previously mentioned performers recorded for Pallás or for Fono-Reyna, who both had a reputation for low prices, suggesting that the wages they paid to their singers were not high either. As multinationals started to record in Spain and well-known singers gradually overcame their reservations against recordings, it understandably became less common for companies to hire individuals with no stage experience and no professional reputation.

Two of the amateur singers who recorded for the *gabinetes* deserve particular attention, given that they are both among those with the most surviving recordings: Blanca del Carmen and Amparo Cardenal. Del Carmen's trajectory is certainly interesting in that she not only recorded for *gabinetes* but also for multinationals in their early visits Spain (Edison Cylinders and Gramophone). This suggests that these multinationals might have given preference to performers with recording experience who knew the processes and the sort of performances that worked well in recorded sound.[68] For the *gabinetes*—and also later on for Gramophone—Del Carmen recorded hits from some of the most successful *género chico* works of recent years, including *La viejecita*, *Gigantes y cabezudos*, and *El tambor de granaderos*. Del Carmen's surviving recordings reveal a reasonably reliable voice, although perhaps not trained (the high notes tend to be in tune, but lack vibrato and roundness), and a rather variable approach to performance and expressiveness: in some of her recordings her diction is clear and expressive, and she makes use of tempo changes and slight portamento to convey an impression of live performance,[69] as many professional *género chico* performers did.[70] In others, however, her performance sounds more mechanical and limited in expressiveness.[71]

Amparo Cardenal, like Sanz Arnal, was a Valencian teenager who allegedly decided against pursuing a stage career for fear of tarnishing her reputation.[72] At Puerto y Novella, she seems to have been the go-to *tiple* for *género chico*, and often recorded in partnership with the previously mentioned Jesús Valiente. She also recorded some *zarzuela grande* numbers, such as the duo from *La tempestad* with Edelmira Guerrero. *Boletín fonográfico* claimed that she was also in demand with Madrid *gabinetes*, but there is no evidence of this either from surviving cylinders or from catalogues and advertisements.[73] From the surviving cylinders, it emerges that Cardenal's voice was pleasant enough to listen to, but not particularly well trained, with unevenness between the registers, some intonation problems and a clumsy approach to coloratura.[74] Her strengths, instead, were in her clear diction and expressive delivery, as well as solid ensemble skills in the cylinders she recorded together with Valiente.[75]

In total, 16 cylinders by Cardenal and seven by Del Carmen have survived. Even though the number is comparatively small, it still amounts to a significant percentage of all surviving recordings of *género chico*. This raises questions about the extent to which these wax cylinders, the earliest recordings in the genre, might give us valuable information about performance practices, given that amateur performers are overrepresented: research into such performance practices has practically been non-existent to date, but future studies will have to engage more fully with the question of whether the recorded samples we have available from these early days are representative of broader tendencies.[76]

Conclusion

Existing scholarship of early recordings has often repeated the idea that the phonograph and gramophone introduced dramatic changes to both performing styles[77] and the music profession in often dramatic ways.[78] What I hope to have provided in this chapter is more detailed evidence of how these changes came about gradually by focusing on a time and a country that have been the object of very little research. These changes took decades to materialize; as a result, it can difficult to identify in the material discussed here any radical, shifts or even clear directions anticipating longer-lasting developments to be effected in the following decades. As has been discussed in earlier chapters, the *gabinetes* industry at this stage was a small, artisanal one, employing a relatively reduced number of singers: it is therefore not likely that the phonograph was seen as a serious threat at this stage to the livelihoods of performers, but at the same time it would not have been seen as a major source of employment for musicians either. At this stage, the reputations of opera and *zarzuela* singers, as is obvious from the specialized publications, were still made on the stage. In this context, and although we lack the crucial first-hand testimonies for singers, it is likely that they regarded it as a curiosity or a further opportunity in the vagaries of trying to put together a career, rather than a make-or-break opportunity, a full-fledged threat or an obligation.

The individual cases discussed in this chapter, nevertheless, suggest that some singers and the *gabinetes* they recorded for were starting to consider some of the issues that the profession would later face on a more widespread level: what did it mean to perform music for recording, as opposed to doing so on stage; how should recordings be fitted within an active stage career. This impinged on recording practices, repertoire decisions, and performance practices, further contributing to shaping the ontologies of recordings that were developing in Spain at the time, as has been discussed in Chapter 3: these did not yet regard recorded sound as separate from its live counterpart, and were still highly dependent on live music and on the rich musical-theatrical culture of *fin-de-siècle* Spain. This relationship with live music, which will be developed further in Chapter 7, must not be regarded as necessarily universal at this stage: even in neighboring Portugal theatrical genres attracted much less recording attention that concert or popular music.[79]

7

Consuming and collecting records
in Spain, 1896–1905

As is the case with the accounts from singers and musicians, the views and
experiences of the first Spaniards to acquire and listen to phonographs and
recordings in the years around 1900 are generally little known to us. In Chapters 1
and 2, I explained how first-hand testimonies from Spaniards who first became
acquainted with recording technologies in the last two decades of the nineteenth century through scientific demonstrations or traveling phonographs are
indeed difficult to find, and the same can be applied to the first customers of
commercial phonography years later. Although *Boletín fonográfico* and *El cardo*
often spoke to the growing community of *aficionados* buying their magazines
and even printed some of their letters asking questions and offering advice on
phonograph-buying and record-making,[1] we must bear in mind that these
were, first and foremost, industry publications that represented the interests of
the *gabinetes* and not necessarily those of their customers. Other accounts regarding the consumption of recording technologies and records can be found
in other newspapers and magazines, and some of them will be discussed in this
chapter—together with a theatrical play in the style of *El fonógrafo ambulante*. As
I have discussed in Chapter 2, these kinds of writings must often be understood
as satire or as making hyperbolic claims for comic effect, but the mocking, critical, or idealized tone they adopt with respect to talking machines is illustrative
in itself of attitudes and discourses around recording technologies at this time.

Next to textual sources, our main source to learn about the first phonograph
owners in Spain is the five substantial surviving collections of wax cylinders
amassed by Spanish collectors in the years around 1900, three of which have
been mentioned already in the last chapter. Frustratingly, none of these five individuals (or indeed any others) left any written accounts as to the circumstances
in which they first became interested in recordings, the reasons why they bought
and collected them, who they bought their cylinders from, what their personal
preferences were, and how they listened to the recordings they owned: whether
alone or in company, whether they were still interested in fidelity first and foremost, or whether they at some point started to develop aesthetic criteria specific
to recording-listening. Nevertheless, the make-up of these collections (in terms
of sheer numbers of recordings, performers, *gabinetes*, genres), as well as some

Inventing the Recording. Eva Moreda Rodríguez, Oxford University Press. © Oxford University Press 2021.
DOI: 10.1093/oso/9780197552063.003.0008

of the scarce biographical details we know about the Spaniards who amassed them allow us to draw some valuable conclusions as to how early phonograph owners consumed recordings, especially when read alongside other sources already discussed in previous chapters. It is significant, in the first place, that these collections were all put together by private citizens or families, and that they only came under the care of professional archives and libraries later in the twentieth or twenty-first century. Indeed, while other countries started institutional collections of recordings (albeit field rather than commercial ones) shortly after 1900,[2] there is no evidence that Spanish institutions ever made a sustained effort at amassing their own collections of recordings at this stage. *El cardo*—likely at the initiative of the Marquis of Alta-Villa, who was a keen singer and singing teacher—suggested in 1900 that the Conservatoire of Madrid should start a library of vocal recordings for teaching and preservation purposes,[3] but the successive *memorias* (end-of-year reports) issued by the institution during these years indicate that the suggestion was not even taken into consideration by the institution's management. Hugens y Acosta claimed in some of its advertisements to be in charge of the sound archive of the Teatro Real in Madrid.[4] The archive, however, has not survived, and I have not been able to find any mentions of it outside Hugens y Acosta's own advertisements, which raises questions as to how substantial it was, or whether it even existed in an organized manner. These instances suggest that some of the *gabinete* owners did indeed realize how significant their recording collections would be decades down the line, but few if any librarians and archivists of their time shared their view.

Among the three collections cited in the last chapter, the most substantial one was that of the Ybarras, one of the most powerful families in Bilbao at the time. The Ybarras significantly contributed to the Basque Country's industrial growth: they were involved in the iron foundry industry since the mid-nineteenth century, and were subsequently one of several similar companies which merged into Altos Hornos de Vizcaya—the largest company in Spain at the time of its foundation in 1902 and subsequently for most of the twentieth century. From the Ybarra collection, 634 wax cylinders were donated to Eresbil (Archivo de la Música Vasca) in the 1980s, of which 551 survive today.[5] Pedro Aznar, a businessman from the town of Barbastro, in Aragón, collected 154 wax cylinders; these are currently held at the Biblioteca Nacional de España,[6] together with cylinders acquired by the Biblioteca elsewhere. Aznar was an acquaintance of prominent *regeneracionista* Joaquín Costa, and both were involved in writing a manifesto from the Cámara Agrícola (Agriculture Chamber) del Alto Aragón in 1898 which is fully in line with *regeneracionista* discourses in favor of the development of the country and particularly its neglected regions.[7] The third collection already introduced in the previous chapter is Antonio Rodríguez Casares's, now held at the Vicente Miralles Segarra museum at the Universitat Politècnica de València. The collection consists of 61 cylinders and a phonograph acquired by Rodríguez Casares in the 1902 exhibition in Paris.[8]

Two further surviving collections have not been discussed thus far because they only contain minimal numbers of cylinders produced in Spain by the *gabinetes*. Leandro Pérez—a printer, shopkeeper, and publisher in Huesca (Aragón), who was also involved in local politics and active as an amateur musician[9]—left a collection of 223 cylinders currently held at the Diputación de Huesca. The collection is unusual in that it consists mostly of Pathé and Lioret cylinders, with just three Spanish ones (two by Puerto y Novella and one by Hugens y Acosta).[10] On the other hand, the collection left by Barcelona-based textile industrialist Ruperto Regordosa and held at the Biblioteca de Catalunya contains only three commercial cylinders (all three by Hugens y Acosta); the remaining 365, all currently held at the Biblioteca de Catalunya under the name Col·lecció Regordosa-Turull, consists of recordings made by Regordosa himself in his own home to a high standard and featuring professional musicians. Most of those were opera singers, but instrumentalists include names as illustrious as Isaac Albéniz's. In the next section, my discussion of domestic phonography consumers in *fin-de-siècle* Spain will proceed under three separate headings: one on the buying practices and preferences of consumers, another one on their listening practices, and a third one on home recording, which was a significant part of the experience of these early collectors. This third section engages with performance practice as a further significant factor that helps us understand home recording procedures; it therefore complements performance practice observations made in Chapter 6, while at the same time expanding the discussion outside commercial environments.

Buying phonographs and recordings

Even the cursory biographical details presented in the preceding section about collectors can provide some useful insights about who bought phonographs and cylinders in Spain at this time and how they might have gone about it. Firstly, all five previously referenced individuals belonged to a socio-economic class we might term as privileged, even though there were significant differences among them individually. Leandro Pérez enjoyed a comfortable lifestyle as well as appreciable notoriety in his small city of Huesca, but his status was not comparable to the Ybarras', who were among the wealthiest, most powerful families in the Basque Country and the whole of Spain. Buying a phonograph and a regular supply of cylinders would not have been, at the time, within reach of a majority of the Spanish population, as suggested by the prices at which phonographs and cylinders were sold at the time. Costs for impressed cylinders have already been discussed in the previous chapter (from 1.5 or 2 *pesetas* for flamenco or traditional music cylinders, to 50 for Napoleón Verger's opera recordings). As for the device itself, Hijos de Blas Cuesta was selling an Edison's Gem phonograph (the cheapest in its catalogue) for 50 *pesetas* around 1900.[11] The same model costed 100 *pesetas* at Hugens y Acosta, but the latter, like other *gabinetes*, included a "starter kit" with the device consisting of a two Bettini diaphragms (one for listening and one for recording), a horn, and other supplies to keep both the phonograph and the cylinders clean.[12] At both *gabinetes*, other models could cost several hundred *pesetas* and even upward of one thousand. These prices would be, if not out of reach, then definitely exorbitant to agricultural and manual workers all over Spain, even taking regional differences into account. According to visiting Austro-Hungarian doctor Philip Hauser, half of *madrileños* survived in 1902 on 60 to 300 *pesetas* per month,[13] whereas in Andalusia a manual worker would earn between 50 and 105 depending on skills and qualification,[14] and a factory worker in the Basque Country could expect to receive between 85 and 115.[15] Moreover, with most working-class families having to dedicate a substantial percentage of their income to basic needs such as rent and nourishment, it is unlikely *gabinetes* would see them as potential customers. As has been discussed in Chapter 2, at this moment in time the working classes would typically rely on *salones fonográficos* and traveling phonographs if they wanted to get acquainted with recording technologies.

Some *gabinetes* made efforts to push their prices down and offer a range of bundles and deals, presumably in an attempt at expanding their own markets. This is the case with Sociedad Anónima Fonográfica selling a phonograph plus six cylinders for 80 *pesetas* as part of their Christmas campaign in 1900[16] and Alvaro Ureña selling in 1903 a kit comprising a phonograph, box, headphones, two diaphragms, and eight cylinders for 100 *pesetas* (which he argued would have costed 200 in normal circumstances).[17] More generally, *El cardo* repeatedly discussed price reductions in the *gabinetes* market under a generally positive light, implying that its editors saw it as a valuable move toward make recording technologies accessible to broader sections of the population.[18]

It is likely, however, that these measures were not introduced with the working classes in mind, but rather sections of the lower middle classes with enough disposable income and an appreciation for science, technology, and/or art (artisans, teachers, office workers). In other countries, we indeed find that recording technologies were at the reach of such classes. Edisonia (the official representative of Edison in London) sold Edison phonographs in 1898 for a minimum of 7 pounds 10 shillings (including diaphragms, horn, two blank cylinders, oiler, and a bottle).[19] Two years later, Pathé was selling its cheapest model in the United Kingdom for 7 pounds 6 schillings.[20] This might have been affordable to some London artisans, although perhaps not so much to other workers elsewhere in the country.[21] With London artisans earning an average of 38 pounds per month already in 1894, this could have been affordable to them, but not so much to, say, Scottish agricultural workers, among whom the most well-off earned only about 51 pounds per year in 1892. In France, Lioret's models sold around 1900 for between 90 and 565 francs[22]—with the cheapest models costing more than half the average salary of a French miner.[23]

Apart from phonograph and recording prices, there would likely be other obstacles preventing working-class families from acquiring their own recording and playback devices. At this stage, being a phonograph user would necessitate a modicum of familiarity with technology and applied science, as was the case with at least four out of the five previously mentioned collectors (we do not have information about Rodríguez Casares' professional or educational background). One of them, Pedro Aznar, was also directly involved in *regeneracionista* initiatives, which could have made him more interested in the discourses about science and technology disseminated by the *gabinetes*, and perhaps in their products too. Central to discourses around recording technologies at this time are also notions of domesticity. As will be discussed later in this chapter, some commentators, as well as some *gabinete* advertisements, placed great emphasis on collective listening together with one's family members or close friends. These discourses would be fully consonant with the notion of the bourgeois household as a center of culture and of physical and moral hygiene, distinct from working class accommodations.[24] They also tied in with what was happening in other countries around the same time: in both Britain and America, for example, the phonograph was accepted as a legitimate presence in the home on the basis of it fitting into existing expectations and practices—from the importance accorded to the visual in the modern home to ideas of music-making in the home.[25] Having a modicum of leisure time was also important: the *gabinetes'* publicity, as well as the phonographic magazines, implied that phonograph owners should devote plenty of time and dedication to acquaint themselves with the workings of the device as well as the range of cylinders available (in contrast, the gramophone was seen as an easy fix whose operation required no particular know-how or talent). Such a hands-on engagement might have been more practicable to the middle and upper classes, and this situates phonography among the "commercial leisure" practices that the Spanish middle and upper classes engaged in at this time, but that remained unaffordable to or unpractical for the working classes for years.[26]

All-in-all, while it is highly likely that the *gabinetes* managed to somewhat democratize access to recording technologies by making prices more accessible to the middle classes, we must bear in mind that, in absolute terms, only a minority of the Spanish population had at the time the means to acquire a phonograph and assorted cylinders. No matter the enthusiasm expressed by *gabinetes* in their trade publications, the impact of commercial phonography at this stage on Spanish society was still relatively limited. This did change to some extent after the widespread adoption of the Gramophone from 1903 onward, with more diverse catalogues, lower prices, and some series—such as Zonophone—specifically targeted to the less wealthy, at the same time that other forms of commercial leisure, such as cinema and sport, became more widely available too.[27] Still, this is not to say that Gramophone fully democratized access to recording technologies in the early twentieth century, since information about the actual circulation of records even at this stage is difficult to obtain and interpret, as is the case with the *gabinetes*, and Gronow has suggested that in its early years Gramophone might have contented themselves with selling a few hundred copies of each record.[28]

The phonograph community in Spain, therefore, was small in numbers at this time, but it is difficult to determine exactly how small. Hugens y Acosta claimed in 1901 that they had two thousand customers, although it is not clear whether this referred to regulars rather than one-off patrons.[29] This is indeed a rather small figure if compared with Spain's total population of 18 million in 1900, and especially if we regard the phonograph in the context of other technological innovations that did cause major transformations in Spanish society around this time: for example, electricity, with Spanish citizens all over the country organizing in electrical societies in the last decades of the nineteenth century,[30] or the cinema, which did cause a veritable revolution in the leisure patterns and preferences of the working classes. Another piece of evidence that suggests that the significance of the phonograph was comparatively limited at the time compared with other technological innovations is the fact that criticism of it was rare. The literary and cultural movements developing around 1898 and concerned with Spain's national identity, values, and future as a country indeed engendered sustained criticism of modern life and the impact of technology on traditional practices and ways of life,[31] but did not engage with recording technologies in great depth. One of the few exceptions comes from the Galician magazine *El Noroeste* in 1905, with the author claiming that both the phonograph and the gramophone brought over "the destruction of the modern home" because "they take everything which is despicable out of words and music without preserving any of their charms."[32]

Apart from helping us ascertain who was buying phonographs and recordings in *fin-de-siècle* Spain, the surviving collections and other documents shed some light on the preferences and buying practices of these individuals, and ultimately on the ontologies of recorded sound they partook in or helped develop, at a time when thousands around the world were engaging contemporaneously in similar practices. Sophie Maisonneuve has pointed out that the figure of the record collector and listener—understood as an individual who fully consumes recorded sound *qua* recorded sound (and not merely as a substitute or proxy for live music) and sees recording technologies as a means of acquiring extensive, encyclopedic knowledge of music (or of a particular genre)—could not fully emerge until the 1920s, with the introduction of electrical recording and the broader accessibility of recordings and devices.[33] However, the Spanish wax cylinder collectors, like their counterparts around the world, offer significant insights as to how initial seeds were planted around 1900 which would then develop into a full-fledged record collecting culture a generation later.

We must start with the caveat, of course, that it is difficult to establish with any degree of certainty how typical or atypical our five individuals were among their fellow recording consumers in Spain. Questions we might ask ourselves when surveying their collections include the following: were most Spaniards attracted by the *gabinetes'* unique catalogues of indigenous genres, such as *zarzuela* and flamenco, like the Ybarras, Aznar, and Rodríguez were?[34] Or were some of them interested in the catalogues of multinationals such as Pathé, which would typically be more extensive and include little-known genres in Spain that audiences could therefore acquaint themselves with, as was the case with Leandro Pérez? How typical was Regordosa in using the phonograph mainly to record his own cylinders, instead of acquiring commercial recordings? To what extent did *gabinetes* shape customers' demand—or did customer preferences shape supply instead?

All of this notwithstanding, the surviving collections of commercial cylinders do reveal relevant information in terms of both personal preferences and broader trends and help us understand how early recording collectors engaged with recorded music. For example, both Aznar and the Ybarras bought from *gabinetes* from all over the country, suggesting that they had a keen interest in diversifying and extending their collection (by contrast, all of Rodríguez Casares's commercial cylinders come from a total of four Madrid *gabinetes*), although there are a few revealing differences in terms of the actual *gabinetes* each of them bought from. Both bought extensively from Madrid and Valencia *gabinetes*, which were the most effective at marketing their products throughout Spain, but they also bought from smaller establishments in their vicinity who are not to be found in any other collections: Aznar bought from La Oriental, the Ybarras from Enrique García and Viuda de Ablanedo, which would confirm that these smaller *gabinetes* might have sold locally rather than nationally. The picture painted by Barcelona cylinders is less conclusive: among these, the Ybarras bought only from the Sociedad Artístico-Fonográfica, but not from Corrons (V. Corrons and then J. Corrons), whose cylinders can be found in the Aznar collection only, even though this was one of the most active *gabinetes* in Catalonia. This confirms evidence advanced in Chapter 4 that the Barcelona *gabinetes* were less successful than their Madrid and Valencia counterparts—with more limited ability to reach outside their city.

As for genre preferences, Rodríguez Casares's favorite music seems to have been flamenco, followed by *género chico*. It is surprising that he did not own any opera recordings, the most popular vocal genre among *gabinetes* at the time. Earlier on I have discussed how the *gabinetes* attempted to find a customer base among the less comfortable middle classes, who would not have typically listened to opera, but rather to indigenous genres: we might hypothesize, on the basis of this and in the absence of biographical information, that he would have been part of such classes rather than of the more elite environments the Ybarras came from. Pedro Aznar seems to have favored *género chico*, with recordings totaling about one third of his collection and followed by instrumental music, but it is more dubious that he had any favorite singers: he owned six by a señorita Martínez[35] who frequently recorded for Corrons, five by Manuel Figuerola, and another five by operatic baritone Marsal, but the rest of his collection is evenly distributed among a multiplicity of performers. On the other hand, the most highly represented works in Aznar's collection generally coincide with recent stage successes (*Gigantes y cabezudos* and *La viejecita*, with four recordings, and *La marcha de Cádiz*, *El tambor de granaderos*, and *La Bohème*, with three recordings each), as well as *Marina* (five recordings), whose iconic status as a Spanish opera has already been discussed in Chapter 6, and *La tempestad* (five recordings), premiered in 1882 and well established by the late nineteenth century as one of the most performed and canonic in the *zarzuela grande* repertoire. Aznar's preferences, therefore, could provide further support to the notion I first introduced in Chapter 6 that recordings functioned at this stage as a memento of live performance, amplifying the notoriety of successful works, while at the same time fulfilling a secondary role with repertoire works. Leandro Pérez's collection, on the other hand, is inevitably shaped by the recordings available in the Pathé catalogue, with examples of *opéra-comique*, *operette*, cabaret, and *varietés*—all of them genres that Pérez is unlikely to have listened to live with some frequency in Huesca. He also acquired, however, from the rather limited selection of *zarzuela* and Spanish song available from Pathé.

The Ybarra collection, being the most numerous and varied of the surviving ones, suggests a more decided ambition to build a collection following certain criteria and preferences—prefiguring in some ways by more than two decades the figure of the record collector/listener as introduced by Maisonneuve previously. Unlike the other collectors, for example, it would appear that the Ybarras had marked preferences in terms of singers, including Rafael Bezares (27), El mochuelo (25), Josefina Huguet (17), Inés Salvador (16), and Lamberto Alonso (11); this can be interpreted an early instance of the cult of specific recorded voices that would develop from 1903–1904 onward, starting with Caruso, and that was actively encouraged by recording labels.[36] The list overlaps with that of musicians of whom the most cylinders have survived, presented in Chapter 6, and is thus a further reminder that our understanding of the *gabinetes*' industry is shaped by the materials found in the surviving collections to extents that are often difficult to separate: indeed, we should consider the possibility that cylinders from the previously mentioned singers survived simply because they were among the Ybarras' favorites and not because they were indeed the most numerous. On the other hand, for at least Huguet, El Mochuelo, and Bezares there is sufficient evidence that they recorded extensively, which opens up the possibility that the Ybarras simply bought cylinders which were readily accessible to them, but were not particularly concerned about the performers they featured. Indeed, details for instrumental recordings are much less conclusive and suggest that the Ybarras might have simply bought what was available to them from neighboring *gabinetes*, as opposed to them developing any strong preferences: they indeed possessed 33 recordings by pianist José Bellver—but Bellver was one of very few pianists to record solo music extensively in Spain at this time, in his capacity as accompanist and musical director for Puerto y Novella; the Ybarras might have simply bought what was available. Many of the pieces acquired by the Ybarras would have simply not been available from other *gabinetes*, so the number of Bellever recordings in the collection does not necessarily mean the Ybarras liked him as a pianist.

Figures concerning repertoire paint a somewhat clearer picture. A small number of operas appear several times, including *La bohème* (14 recordings),[37] *Marina* (11), *Carmen* (10), *Cavalleria rusticana* and *Aida* (8 each), *Faust, Lohengrin, Mignon,* and *Rigoletto* (7 each), *L'Africaine, La Dolores, La favorita, Pagliacci,* and *Un ballo in maschera* (6 each), and *La Gioconda, Lucia di Lammermoor* (5). Many of these—namely Verdi's operas, as well as *Cavalleria, Pagliacci,* and *La Gioconda*—were relatively recent successes, whereas others (especially Italian bel canto as well as *Marina* and *La Dolores*) were well established in the repertoire. A smaller number of *género chico* works also have appreciable numbers of recordings, including *Gigantes y cabezudos* (7), *El dúo de la Africana* (6), and *El cabo primero* (5)—all recent successes of the genre, even though other equally successful works were less prominently represented (e.g., just one recording for *La viejecita*). This further suggests that the Ybarras regarded phonograph recordings as a medium to preserve and revive some of the experiences they would have had with live music—as prominent members of the Bilbao bourgeoisie, they would have been regulars at the Teatro Arriaga—rather than as a means of becoming acquainted with music they might not have been familiar with.

Particularly intriguing is the fact that the Ybarra family held multiple copies of some arias or romanzas; in most cases, these would be by different singers, although with a few exceptions.[38] Most of the duplicated arias belong to the lyric and spinto tenor repertoire ("Celeste Aida," "Vesti la giubba," "O Paradis"), of which the Ybarras tended to hold two or three copies by different tenors, most notably Bezares, Alonso, Julián Biel, and Ignacio Varela. This is consonant with the popularity and status that tenor repertoire had on stage at the time, but at the same time suggests that certain medium-specific ways of acquiring and listening to recordings (as opposed to live music) were starting to develop at this stage. Indeed, although we cannot know for sure, the Ybarras might have listened to different versions of the same aria for the purposes of establishing more-or-less detailed comparisons that would have not been possible in a live context.

Learning new ways of listening

Engaging in detailed comparisons of recordings, as the Ybarras might have done, was likely just one of several new ways of listening to music that first appeared during the era of the *gabinetes* and would in successive years originate more widespread practices we now often take for granted in our dealings with re-corded music. Some of these new ways of listening might have been simply the natural outcome of the process of becoming acquainted to the new medium: in-deed, the *gabinetes'* own publicity suggests that frequent listeners and buyers be-came more discerning and selective regarding the quality of the cylinders, and, when making buying decisions, they gave importance to fidelity and a satisfac-tory listening experience.[39] The experience of these early recording listeners must have also been colored by the fact that cuts were routinely introduced into pieces so that they could fit into a three-minute cylinder. Changes in the music could also reflect the financial limitations faced by *gabinetes* at this time: for ex-ample, in her very expressive recording of "Quando m'en vo" for Blas Cuesta, soprano María Vendrell does sing the short dialogue with Marcello toward the end—but the *gabinete* did not hire a Marcello and so Vendrell simply sings her lines to no one.[40] More rarely, short arias could be elongated rather than short-ened, as is the case with one of Huguet's recordings for "Saper vorreste" for Viuda de Aramburo, in which the second and last stanza is repeated no less than three times, presumably to fill the remaining space in the cylinder with sound.[41]

Whereas the phonograph and gramophone would have enabled solitary lis-tening to become a widespread practice for the first time in history around these years, evidence suggests that many Spaniards still chose to listen to recordings with their family and friends (which also ties in with the notation that solitary listening only really took off at the same time as the figure of the record/collector listener developed in the 1920s and 1930s). This preference for group listening can be understood with reference to both domestic music-making practices and to the traveling phonograph and *salón* demonstrations, in which the ability to compare and comment responses and impressions with others would have likely been an important part of the experience. It certainly was elsewhere, with a British catalogue from Edisonia (the official distributor for Edison phonographs and cylinders) stating in 1898:

> You must have friends. And they and you sometimes meet. If you are the host you can entertain them more perfectly by the means of this versatile instrument than in any other manner. You can make it sing to them, either serious, senti-mental or comic songs, recite to them, repeat the speeches of celebrities, play the music of any instrument, perfectly reproduce a full band or orchestra, and, greatest of marvels, you can get your friends to speak into the instrument.[42]

The gramophone, on the other hand, was also seen in its early years as ideal to facilitate collective listening. "It would be practically an impossibility for an-yone to give such a musical treat to their friends by any other means than The Gramophone," stated a British Gramophone advertisement in 1904.[43] The same advertisement suggested a program of specific Gramophone records to be played back to one's circle, in what constitutes an early example of recording labels shaping ideas of taste and the musical canon that would develop in later decades.[44]

Social or collective listening was not such a central feature in the advertisements of the *gabinetes* as it was in other countries (and, in any case, we should not assume that consumers would always adopt and follow the listening practices suggested by the publicity), but other sources suggest that Spaniards too might have also engaged in this practice. As with some of the sources illus-trating early attitudes to recording technologies throughout the 1880s and in the era of traveling phonographs, some detail comes here from theatrical and satir-ical sources, such as *Los cuatro trapos*, a 1908 *sainete* with music by Luis Foglietti and text by Antonio F. Lepiña and Antonio Planiol. The play features a gramo-phone played in a café, with the audience offering comments on the quality of the performers and the recording. These comments, which are overly admirative or dismissive, are obviously exaggerated for comic effect, but it is not a stretch of the imagination to picture similar dynamics taking place in real-life collective listening sessions. The article I discussed earlier in *El Noroeste* also describes a collective listening session among friends, in which the attendees became disap-pointed at the poor quality of the sounds of the gramophone.[45] Well-known play-wright and Nobel Prize winner Jacinto Benavente wrote an idealized describing a high society reunion, which we can presume would draw on relatively wide-spread listening practices among certain echelons of Spanish society.[46] In Benavente's account—which could be real or fictionalized—a phonograph ("the most fashionable toy") is exhibited and played back: "the marquis plays a bell and a man servant appears, who is then duly instructed. He leaves and then comes back with a phonograph and a box of cylinders." Benavente then describes the marquis playing back a political speech followed by a French song; the civilized and friendly nature of the gathering is emphasized (and idealized) throughout, as in other journalistic accounts of real-life meetings where the phonograph was often exhibited alongside instances of live performance.[47]

Doing it yourself: Self-recording in the *gabinetes* era

Chapter 3 discussed how the *gabinetes'* publicity strategy focused rather extensively on the home recording possibilities (of both music and spoken word) offered by phonographs. This was often presented as a key advantage of the phonograph over its direct competitor, the gramophone.[48] Several phonograph owners seem indeed to have taken up the invitation, with *El cardo* and, to a more significant extent, *Boletín fonográfico* printing letters from readers who either asked technical questions about self-recording for the magazine's staff to answer, or offered technical advice themselves.[49] A. Marín conceded that some home-made recordings could reach very high quality standards on a par with commercial cylinders, with the human voice being reproduced "with clarity, with no stridency or gargling,"[50] and magazines themselves encouraged home recording. A contest was organized in January 1900 to celebrate the launch of *Boletín fonográfico*, where amateurs were invited to record three set pieces ("Voi lo sapete," Gounod's "Ave Maria," plus a free choice spoken word cylinder), with the winners to be chosen by the directors of the Valencia *gabinetes* and music critics.[51]

Some of these recordings might have extended beyond what were strictly the boundaries of the home, thus tying in with some of the collective listening practices from earlier eras: *Boletín fonográfico* reported that a reader from A Coruña had sent a recording of a choral society in A Coruña—perhaps El Eco, a well-established choral society in Galicia.[52] The surviving cylinders also provide evidence that some recordings were made at social organizations and circles rather than in the home: the Biblioteca Nacional de España holds a cylinder labeled as Sociedad Filarmónica de Medina containing the song "A Granada," by an unknown performer,[53] and Emilio Cabello's recording of "Cuplés de Teatro" is labeled as Círculo de la Amistad in Barbastro.[54] Given that the latter cylinder was part of Pedro Aznar's collection, we might hypothesize that he, as a resident of Barbastro, might have been involved in the making of the recording.

Evidence from the surviving collections confirms that self-recording was likely a significant part of the experiences of early users of recording technologies. The most conspicuous case is, of course, Regordosa, whose collection will be discussed in more detail subsequently, but at least three out of the other four engaged in self-recording to some extent. There is no firm evidence that the Ybarras ever did, but the collection's inventory includes details of half a dozen cylinders by "amateur" performers (señor Villabella, Satur), currently non-digitized; this might be an indication that these were indeed made in their own home. Pedro Aznar's collection includes a small number of self-recorded cylinders of instrumental music[55] as well as vocal genres (an opera aria by his relative Ascensión Marro,[56] flamenco by an Isabel Gómez[57]) and spoken word cylinders. Rodríguez Casares's collection does similarly include a few cylinders of unaccompanied singing (mostly *zarzuela* arias), presumably involving members of his family or close friends (as they are only referred to by their first names), as did Leandro Pérez's: it is indeed the fact that they were unaccompanied that suggests that they were not professionally produced. All of the previously mentioned recordings undoubtedly have cultural and historical interest: they indeed prove that self-recording practices played a significant role in the early days of the phonograph in Spain, and further underline the social and collective nature of recording-listening and recording-making at the time, with most of the preceding recordings featuring family members or otherwise people from the collector's closest circles. From a musical and performance practice point of view, though, they might be less significant, as the performers featured in those were amateur. In this, they differ from the eight cylinders Pérez made of violinist José (Pepito) Porta and, most notably, from the Regordosa collection. The former consisted of eight cylinders that Pérez himself accompanied at the piano[58] at a party his own house in 1907.[59] Porta, originally from the Aragonese town of Sariñena, was then 17 and had already acquired recognition in his native region as a child prodigy; he had also received a number of scholarships from the provincial government to study abroad.[60] He would then become a violin teacher at the Conservatoire of Lausanne (where he died in 1929), and took part in the premiere of the trio version of Stravinsky's *Histoire du soldat*, together with Spanish pianist José Iturbi. Joan Chic has pointed out that Porta's playing in these cylinders is "very far removed from the false stereotype of the early twentieth-century violinist, with a velvety, sweet sound and somewhat lacking technique,"[61] and that it features sparse use of portamenti and glissandi (which we tend to associate too with these performing styles); vibrato is similarly, scarce, which is more in line with our expectations.

Pérez's recordings of Costa are indeed an important source for the study of historical violin performance practice as well as home recording culture in early twentieth-century Spain. Their importance, however, is surpassed by Regordosa's collection—and this not only in the Spanish context, but on an international level too. Although the total number of cylinders in the collection nears four hundred, the number of individual musical pieces is more reduced (260, owing to the fact that some cylinders contain spoken word and not music, and that some of the pieces are split between two, three, or four cylinders). Still, this clearly puts it above other non-commercial recording collections from the same era: Julius Block, a Russian business and phonograph enthusiast who recorded such luminaries as Sergey Taneyev, Anton Arensky, Paul Pabst, and a young Jascha Heifetz in his home, only left behind about one hundred musical recordings;[62] and a similar number resulted from the efforts of Lionel Mapleson—a librarian at the Metropolitan Theatre in New York who recorded singers live on the Met stage for no commercial purposes.[63] The Regordosa collection is significant not only in numbers, but also in terms of the performers it features: indeed, even though the number of instrumental recordings is limited (19), they include an improvisation by noted composer and pianist Isaac Albéniz, as well as recordings by key figures in Spanish and Catalan musical life of the time, such as pianists Joaquim Malats and Frank Marshall and clarinetist Josep Nori. The majority of the recordings is of vocal music,[64] and prominent names are represented here as well. Indeed, several of the singers who recorded for Regordosa also did so for multinationals such as Fonotipia and Victor; most were Spanish (Ramón Blanchart, Josefina Huguet, Ramona Galán, Avelina Carrera, Andrés Perelló de Segurola, José Torres de Luna), as well as a few Italians (Edoardo Garbin, Mario Sammarco). Regordosa also recorded the famed Orfeó Catalá, a Barcelona choral society which also recorded extensively for multinationals (and continues to do so up to the present), as well as solo singers from the Orfeó entourage (Emerenciana Wehrle, Amparo Viñas) who held prominent roles in Barcelona musical life and particularly those aspects of it connected to the nascent Catalan musical nationalism. This makes the Regordosa collection a key source for the study of early twentieth-century performance practices in vocal genres, particularly opera, whose full potential has not been explored so far. Perhaps even more interestingly, the collection can illuminate our understanding of home recording practices during the era of the *gabinetes*, both in Spain but also worldwide, given that it is one of the most significant collections of home recordings to have survived. Difficulties for the study of the collection include fact that no visual or textual sources pertinent to the collection have survived which could supplement or support what can be inferred from the cylinders themselves. Similarly, we must bear in mind that Regordosa was probably unusual in his dedication to home recordings in terms of the sheer amount he produced, the well-known names he attracted, and the technical standards to which he made his recordings (on which more later), and it is more likely that the self-recording experiments of an average phonograph owner in 1900s Spain would rather resemble the more modest endeavors we observe in the collections of Rodríguez Casares, Aznar, and Pérez.

We do not know, in the first place, what motivated Regordosa to start his collection and increment it until it got to several hundred items. The scarce biographical records we have about Regordosa, though, reveal that he was rather typical of those who first engaged with commercial phonography in Spain: as a textile industrialist, he would have reasonable familiarity with new technologies, and he was also involved in factory and business owners organizations in Barcelona,[65] so it is likely that he would have been familiar with discourses that placed entrepreneurship, trade, and technology at the center of national rebuilding—be it for Spain or Catalonia—which, as has been discussed in Chapter 3, was also at the basis of the publicity strategy of the Madrid *gabinetes*. Consonant with his bourgeois credentials, he was also a regular at the Liceu.[66] It might have indeed been at the Liceu that he scouted and perhaps first contacted singers he wished to record, as an overwhelming majority indeed sang at the Barcelona opera theater between 1898 (the date of manufacture of Regordosa's oldest phonograph, out of a total of four he eventually acquired) and 1918 (the date of Regordosa's death). This suggests, yet again, that Regordosa's engagement with recorded music was closely connected to his experience of live music, sometimes recording singers in arias from roles they had sung at the Liceu. One of the most interesting instances in this respect concerns soprano Avelina Carrera's recording of "Leise, leise" from *Der Freischütz*. Carrera sang the role of Agathe in May 1903,[67] which made it the first time in 17 years that the opera was performed at the Liceu. None of the Regordosa cylinders contain indications of the date where they were made, but we might consider the possibility that Regordosa made the recording shortly after Carrera's appearance, as a memento of the live performance.

Several of Regordosa's recordings contain spoken introductions by the singers themselves or short conversations with Regordosa; from this, we can infer that the industrialist might have been friendly with some of the singers he recorded (such as Carrera herself and Ramona Galán) and that the atmosphere in which recordings were made was relaxed compared to the *gabinetes' salones de impresión*, although, as will be discussed later, this was not in detriment of quality standards. Whereas several of Regordosa's singers did record for the *gabinetes* or for multinationals, others never did, such as Lina Cassandro and Amalia de Roma, and it might have been this relaxed atmosphere, as well as Regordosa's status as a respected industrialist and the fact that the recordings were not meant to be released commercially, that encouraged those singers to take the plunge. With Regordosa's and the *gabinetes'* recordings not being dated, it is impossible to confirm the hypothesis that some singers might have familiarized themselves with the recording process at Regordosa's home and then went on to record professionally, or that, conversely, Regordosa might have targeted singers with previous experience to improve his odds of obtaining good results.

Regordosa, indeed, seems to have been concerned with obtaining good results—meaning the achievement of artistic and technical quality in his recordings on a par with the standards achieved by the *gabinetes* at that time, and, to a great extent, he was successful. Although it is likely that he developed his expertise by trial-and-error—as did all those who tinkered with phonographs at the time, professional and amateur alike—there is no evidence that he had any help along the process. There is no trace in Regordosa's collection of any contacts he might have had with the Barcelona *gabinetes* (he could have acquired his Hugens y Acosta recordings by mail order), and there is no evidence either that he subscribed to *El cardo* or *Boletín fonográfico* and might have therefore benefitted from the practical tips provided there. Nevertheless, the extent to which Regordosa's own recordings adhere to the *gabinetes'* conventions and standards suggests that he might have been more familiar with professional recordings than the collection suggests. For example, many of the pieces were cut to allow them to fit into cylinders, although others were recorded in their entirety and split between two, three, or four cylinders—unlike the *gabinetes*, who did this only very exceptionally. Similarly, an overwhelming majority of the cylinders contain a spoken announcement at the beginning; the male voice that speaks in most of them is likely to be Regordosa himself, but a female voice speaks in some of them too (likely to be Regordosa's wife), and so do two other male voices, which suggests that Regordosa could have shared his recording sessions with at least a small group of friends or relatives. Interestingly, Regordosa put on an Andalusian accent to introduce the flamenco recordings, which was also common practice in some of commercial cylinders. The piano accompaniment in the recordings is competent and of a professional standard; in some of the recordings the accompanist is named as señor Puig, but there is no indication of who he was or whether he played in all recordings. Despite the relaxed atmosphere, there is no background noise or indeed any indication that the recording sessions would have taken place in the context of a party—as did Pérez's recordings of Porta and several of Block's recordings.

Some of the singers' choices of repertoire also point toward a relaxed atmosphere, and otherwise constitute an interesting point of comparison to the *gabinete* practices that have been discussed in Chapter 6. As it is to be expected, singers recorded mostly within their Fach or specialization, that is, what they would have normally sung on stage and become known for; this, of course, also overlaps substantially with the repertoires these singers would have recorded commercially, with some exceptions as discussed in Chapter 6. Nevertheless, it is likely that the atmosphere of Regordosa's sessions, as well as the fact that these recordings were not meant to be released commercially, led some singers to deviate from their usual stage repertoire even further, resulting in a small collection of rarities. Of these, the presence of Catalan-language song is one of the most interesting areas, both musically and culturally: indeed, as has been discussed in Chapter 4, there is no evidence that the Barcelona *gabinetes* produced any Catalan-language recordings, which turns the Regordosa cylinders into some of the earliest recorded examples of this repertoire and at the same time opens up questions about the role of recording technologies in establishing and disseminating the nationalist music movement which was taking shape at the time.

Regordosa recorded four arrangements of Catalan traditional songs (both for choir and for a soloist with piano accompaniment), and 18 modern Catalan-language songs inspired by traditional music (14 for soloist and piano and four choral pieces), by composers such as Anselm Maria Clavé, Lluís Millet, Pep Ventura, and Josep Borrás de Palau. Also part of the collection are two numbers from the Catalan-language *zarzuela* (*sarsuela*) *Les barraques*, by Vicente Peydró. Some of these were recordings featured, as has been indicated earlier, by the Orfeó Català or solo singers from its environment (Wehrle, Viñas), but a sizeable number were recorded by opera and *zarzuela* singers of Catalan or Valencian origin who would thus be familiar with the Catalan language, even though they would not normally perform this repertoire in public or record it commercially (Josefina Huguet, Concha Dahlander, Andrés Perelló de Segurola, José Sigler, Marina Gurina). It is interesting that Regordosa never recorded *sardana*; this might be because, as has been discussed in Chapter 4, the genre was then just starting to spread outside its native area of Empordá and becoming transformed into a symbol of Catalan identity, so Regordosa might not have been overly familiar with it yet. Otherwise, he might have also been put off by the logistics of engaging and recording a full *sardana* ensemble, rather than a soloist with piano.

Another interesting instance of a singer performing outside her core repertoire for Regordosa is Marina Gurina's (a *zarzuela tiple*) recording of "Voi lo sapete."[68] Gurina was, in her time, regarded as a good singer with the vocal skills to perform demanding *zarzuela grande* roles. Her recording of "Voi lo sapete," though, reveals obvious vocal shortcomings, with high notes sounding particularly unstable. Some of her *zarzuela* recordings for Regordosa show similar issues,[69] while others do not,[70] suggesting that these problems were not necessarily caused by unfamiliarity or lack of ease with the operatic repertoire: it might be, for example, that Gurina's singing simply became more reliable as she grew more comfortable with the recording process at Regordosa's home and at *gabinetes*.

More generally, Gurina's performing skills and her sense of the text in this re-cording do result in an expressive, intelligent performance of "Voi lo sapete," and this might help us understand better the sorts of skills (in acting, singing, musi-cianship, and text delivery) that *zarzuela* singers were expected to display, and how they juggled them to produce successful performances both on stage and, later on, in recordings.

Josefina Huguet, who was introduced in Chapter 6, recorded extensively for Regordosa, mainly in her usual soprano coloratura repertoire, including both bel canto and French *grand-opéra*. She also recorded Catalan-language songs, but her most surprising contribution is perhaps a recording of Schubert's "Ständchen"[71]—a rarity both within Regordosa's and the *gabinetes'* collections, since the German Lied was practically never recorded at all in Spain at this time. Huguet's recording is certainly very different from what we might con-sider a good Lied performance today—but it is different, too, from Julia Culp's 1914 recording for Victor, one of the earliest to be documented.[72] Huguet sings the song considerably slower than Culp (crotchet = 55, to Culp's 73), which forces her to skip the repetition to keep the song under three minutes, and she certainly sounds out of her depth in the recording: her middle-low and middle-high register are uneven, she cannot capitalize on the brightness of her high register, and the highest note in the song (a F5 in bar 23) sounds out of tune. Nevertheless, Huget clearly attempts to be expressive by adopting strat-egies she used in coloratura repertoire—subtle ornaments, portamento (es-pecially at the end of phrases), and, in the opening phrase, a generous use of *tempo rubato*. Perhaps most surprisingly, she sings bars 50–52 an octave higher than written and interpolates an A5 in the last two bars). This rather unconventional but nevertheless interesting recording forces us to con-sider what Regordosa's collection reveals about non-standard performing practices: Huguet's "Ständchen" should probably not be regarded as a model of Lied performance practice, but it certainly illustrates how performers might have imported conventions from a genre or style to others—as is the case too with *zarzuela* and opera.

The end of the phonograph era

Whereas we can presume that the figure of the Spanish phonograph aficionado and cylinder collector appeared at the same time as the *gabinetes* started to market their products, it is more difficult to ascertain when the community who had gathered around the *gabinetes* themselves and around *Boletín fonográfico* and *El cardo* might have disappeared. The rise of the gramophone between 1903 and 1905 at the expense of the phonograph probably had a significant impact, with many of the earlier consumers moving from the former to the latter without much trauma. In fact, whereas in other countries wax cylinders and gramophone discs coexisted for almost three decades, in Spain the transition from one playback device to another seems to have been completed relatively quickly. Gelatt claims that, on a global scale, the cylinder as a medium started to decline from 1901 onward,[73] but the decline was long and convoluted, with Edison periodically launching new innovations aimed at reviving the phonograph over the gramophone. Phonographs and graphophones were still being advertised between 1905 and 1908 in the United Kingdom and the United States,[74] and Read and Welch indeed situate the peak of the wax cylinder in the former country in 1907.[75] In 1913, Edison launched a new type of cylinder, the blue Amberol, in a new operatic series featuring tenor Alessandro Bonci, among others.[76] A devoted group of phonograph fans, particularly in the Deep South, continued buying primarily cylinders until production ceased with Edison's retirement from the recording business in 1929.[77] In Spain, though, given that the phonograph had been primarily introduced and marketed by the *gabinetes* and not by Edison himself, there is no evidence that cylinders were sold on a commercial scale once the *gabinetes* closed down or became Gramophone resellers. Some advertisements for second-hand phonographs and cylinders can still be found in newspapers, but these would not have constituted an industry.

Nevertheless, given that the transition from phonograph to gramophone excited some opposition not only from the *gabinetes* themselves but also from users,[78] and given that one of the phonograph's appeal was in producing one's own recordings, we might hypothesize that at least some phonograph owners would continue to use their device in their own home for years after the closure of the *gabinetes*, perhaps buying cylinders (blank or impressed) from abroad. Regordosa, again, provides an example of this. Even though none of his recordings are dated, he must have been recording at least as late as 1906, given that this was the year when Amadeo Vives's concert song *La riojanica* was premiered. We must consider the possibility that Regordosa indeed kept making recordings until his death in 1918, although it seems more plausible that most of them date from before 1909 or 1910: indeed, some of the singers he recorded (José Sigler, Concha Dahlander) retired or died before these dates and the dates of the visits of singers to the Liceu suggest that all of them were in Barcelona at least once before 1910, whereas none of them visited the city *only* after this date. Regordosa, therefore, must have remained active making his own recordings for at least a few years after the *gabinetes'* closure, seemingly impervious to the new allure of the gramophone which was captivating Spaniards; perhaps others did too.

Conclusion: From national to global

From their inception and throughout their early decades, recording technologies existed in something of a paradox when it came to positioning themselves alongside spatial axes. On the one hand, phonographs and recordings were widely regarded as transnational and global: even in the years before recorded sound developed technologically and commercially, accounts of Edison's invention awarded great emphasis to the fact that the phonograph would allow sound (and especially the human voice, whether in singing or in speech) to travel across cities, countries, and continents, and also across eras and generations. Edison himself became a worldwide celebrity, hailed as one of the great inventors of the nineteenth century. And, once recorded sound became commercially viable, companies—Edison and its direct competitor Gramophone first, then Victor, Columbia, and Lindström, as well as Pathé, among others—set out to both secure markets for their products overseas and engage some of the best performers active at the national or local levels elsewhere, often engaging in fierce competition, commercial and legal, with each other and with smaller labels. The fact that by 1914 the young recording industry was transnational and global at its very core is obvious from the fact that it entered into its first crisis when the First World War substantially curtailed the circulation of devices, recordings, and musicians.[1]

At the same time, though, the processes that took recording technologies from mere a scientific potentiality to commercial and cultural artifact in the course of a few decades also emerged from substantial engagement with the national, the regional, the local, and the ultra-local, and this often took experimental, quasi-haphazard ways that are not easy to reconcile with the idea of Edison, Berliner, and a few other masterminds orchestrating a global takeover from their offices in the United States, Berlin, or Paris. From a relatively early stage, inventors everywhere patented accessories and improvements to the phonograph, and sometimes even ventured to build their own talking machines and make their own recordings, whereas entrepreneurs from a variety of backgrounds within the broad banner of applied science and technology sought to develop commercial strategies to earn money out of these. In doing so, they had to work within their own sets of regulations and practices (which were dictated by national, regional, or municipal bodies). But we should not think of these local agents and entrepreneurs as being always inimical to multinationals: indeed, several multinationals soon realized the advantages of decentralization and they built

Inventing the Recording. Eva Moreda Rodríguez, Oxford University Press. © Oxford University Press 2021.
DOI: 10.1093/oso/9780197552063.003.0009

networks, of different types, with local business and individuals who could open up access to local markets (both of customers and of musicians); this practice was still widespread in the 1920s.[2]

At the same time, though, using the categories of the global, the transnational, the local, the ultra-local, and others in the period at hand necessitates a flexible approach, rather than considering them fixed spatial planes: indeed, the decades around 1900 were one of the periods in modern history which saw the most dramatic transformations in terms of how these and other categories were thought of. Regarding the global and transnational levels, for example, the early decades of recording technologies coincided with the last of colonialism, and also with exponential developments in transport, communications, and other technologies that dramatically redefined distances in complex ways: getting from certain parts of the world to others now took considerably less time than it had a mere twenty years earlier, and economies became increasingly interconnected, but the new era also opened enormous gulfs, literal and figurative, between different parts of the world. The boundaries of the nation-state (and, consequently, the very idea of the national) were thus being redefined from the outside, but, importantly, also from within: nation-states emerged from the unification of territories that had been politically independent thus far, whereas others became independent when detaching themselves from larger states or empires. In other cases, such as Spain itself, territorial borders might not have changed much in a literal way, but ideas of the nation-state certainly did as nationalist and independence movements developed in some of their territories. This, in turn, sometimes led to the development of the regional as a separate category to allow for a certain degree of diversity and difference while keeping regions strongly constrained within the boundaries of the nation-state. Finally, notions of locality were developing and changing too, sometimes in opposition or as a reaction to some of the previously mentioned developments in the transnational/global or the national, but also as a consequence of other developments such as the growth and transformation of the cities, stimulating new types of discourses about the urban space and urban identities. As I hope I have demonstrated throughout this book (for example, in my discussion about traveling phonographs and discourses of mobility, as well as in my considerations about *gabinetes* and urban space in Madrid and Barcelona), recording technologies were not always passively shaped by all of these developments: they were also an active part of it, trespassing boundaries (literal or figurative) and redefining spaces.

But, beyond generalizations and generic admonitions to consider the interconnected roles of the transnational/global and the national/regional/local, what lessons can be learned from the Spanish case study that might have further relevance elsewhere? In writing about early recordings and early recording technologies, how can we make sure that we are moving productively between these different lenses and perspectives to gain as nuanced an understanding as possible of these complex realities, without dismissing difference as simply local particularities with no further relevance? Alternatively, if we choose to stay within one of these perspectives—which is an equally legitimate approach to take—how can we do so while acknowledging that our chosen perspective will occasionally interface with others on a different plane?

Although I certainly do not claim universal knowledge of the particulars of the history of early recording technologies in each country or region, I would like to highlight four such lessons emerging from the present book: four areas in which the importance of the local and the national for the shaping of the idea of the recording was particularly felt in Spain and can therefore be reasonably expected that it will be significant in other contexts too. I contend, at the same time, that these four areas, where certainly emerging to an extent from previous research, have not always been sufficiently accounted for or could be given further critical consideration.

First of all, it emerges from the Spanish case study that the role of Edison, his companies, and other recording multinationals in crafting and spreading the idea of the recording as a concept and setting the foundations of the recording industry might need to be further re-evaluated and contextualized. It is, of course, understandable that a great deal of the research on the origins of the recording industry focuses on multinational companies: in many cases, they provide direct continuity between the early years of recorded sound and modern times. Moreover, their recordings and records about them are relatively easy to track down and interpret, with some of these companies having their own archives, such as the Edison Papers at Rutgers University and the EMI Archive Trust, while on the other hand (as I have discussed in the context of the *gabinetes*) sources and recordings for smaller companies had fewer chances of being preserved throughout the decades. In these narratives, smaller, often local companies are often mentioned in passing, and only to the extent to which they competed with multinationals before they were absorbed by them or forced to close down, often through legal action, which can sometimes introduce a distinct (if perhaps non-intentional) "survival-of-the-fittest" tone to the narratives of early recorded music. Even though we must acknowledge that companies with transnational ambitions sprouted very early in the process that took us from recorded sound to the recording and became dominant within a very short period of time, we should not regard smaller labels and individuals working on their own as inevitably doomed from the start or as relics from parts of the world that even the multinationals initially overlooked. Instead, I argue that it is more productive to study and ascertain the role of these smaller companies and agents in shaping practices and discourses in certain contexts, as I have done here from the *gabinetes*, zeroing in on a period that is often glossed over in histories of recording in national contexts: sometimes simply on a local level, but sometimes more broadly. It can also be productive to regard them as the first pioneering specimens of "do-it-yourself," low-cost, or small-scale strands of recording-making that have been a constant—even if not always with the same degree of visibility—in the history of recording technologies: let us think, for example, of the cassette tape or, more recently, of the latest developments in digital technologies which allow a broader range of individuals to make their own recordings and, in some cases, market them.

A second area of concern regards the national and local characteristics that impinged discourses around modernity in the period at hand. Recording technologies, *qua* technologies and products of applied science, were undoubtedly regarded as products, carriers, and symbols of modernity, and modernity tended to be connected too with discourses of the transnational and the global. A modern world, in the period at hand, was a world that was more interconnected and in which distances were shorter. At the same time, though, Spain was by no means the only country where these shared concerns about modernity intersected with its own national concerns and anxieties—past, present, and future: modernity, while being significantly about the international, the transnational, the interconnected, often tended to acquire distinctly national or local flavors. These too had an influence on how recording technologies were thought of even from the moment when they were no more than a possibility: how their potential was envisaged and which particular qualities and characteristics were emphasized and which others were paid less attention, according to the sorts of outlets and possibilities they offered to national anxieties.

Discourses of modernity, with their national underpinnings, including discourses around science and technology, are in turn connected to the third area I think emerges clearly from this book: namely, that the history of early recording technologies is also a history of possibilities. Indeed, in these early years a multiplicity of uses for recording technologies, as well as practices attached to them, were imagined, written, and spoken about; of these, only a few ended up materializing. Of these, some have stayed almost unchanged until our days, others transformed, others disappeared without a trace: let us think, for example, of the uses that Edison originally envisaged for the phonograph, primarily as a dictation and secretarial machine, with only minimal importance being accorded to musical entertainment. Even a cursory look at those speculations, uses, and practices should convince us that the reasons why some possibilities that were opened up by recording technologies ended up materializing whereas others did not do not simply concern technological or commercial viability, but are also deeply connected to cultural matters—and these, in turn, would often be shaped by national, regional, or local cultural specificities: it was not a matter of technological inevitability or technological Darwinism where only the fittest possible devices and practices evolved and survived. Cultural matters are relevant too when considering histories and genealogies of these practices, as well as continuities and discontinuities: I have explained, for example, that some of the ideas and practices connected to traveling phonographs left their mark when phonographs went into the domestic sphere a few years later. While it is true that it makes sense to tell the history of early phonography synchronically, focusing on travels, mobility, and exchanges between one country and another,[3] it also makes sense to do so diachronically, considering how the emergence of a particular practice or particularity in a specific geography, no matter how unique or abnormal in an international context, might have precedents within the same country and be broadly informed by ideas of modernity and technology and science but also by past uses given to technologies (of sound or otherwise). In the same way as Sterne and Gitelman[4] have demonstrated that there is a genealogy of ideas and practices behind the emergence of recording technologies themselves, we should consider the idea that these genealogies sometimes happen within the borders of the nation-state or even of a city, for a number of reasons having to do with policies and regulations, but also with discourses concerning science, modernity, and technology that are strongly shaped by ideas of the national, regional, or local.

The fourth and final area concerns the music and repertoires that were recorded in these early years of the phonograph. This is perhaps what first attracted me, as a musicologist, to the study of early phonography, although in the meanwhile I had numerous opportunities to realize, as I hope is obvious from this book, that early recording experiments were not all about music. Recording technologies—mostly through the relatively well-studied phenomenon of "ethnic records" of some of the multinationals—undoubtedly played a role in making some musical genres better known outside its place of origin, sometimes deriving in interesting dynamics:[5] we could cite here the case of flamenco, which encountered recordings relatively early in its development and has ever since been marketed internationally, often under the assumption of being exotic. Recording technologies too played a role in homogenizing of taste and musical practices, making popular music quasi-synonymous with English-speaking popular music. But, other than that and as I hope this book has shown (especially Chapter 6), local or national repertoires, especially those with strong theatrical or stage traditions, also contributed to shaping recording practices and ontologies of recording technologies in particular ways, rather than recording technologies being used to passively record what was already there.

Notes

Introduction

1. Sophie Maisonneuve, "De la machine parlante au disque. Une innovation technique, commerciale et culturelle," *Vingtième siècle. Revue d'histoire*, 92 (2006), 17–31.
2. Peter Martland, *Recording History. The British Record Industry, 1888–1931* (Lanham, Toronto, and Plymouth, UK: The Scarecrow Press, 2013); Marina Cañardo, *Fábricas de músicas. Comienzos de la industria discográfica en Argentina (1919–1930)* (Buenos Aires: Gourmet Ediciones, 2017).
3. Pekka Gronow and Ilpo Saunio, *An International History of the Recording Industry* (London and New York: Cassell, 1999).
4. Leonor Losa, *Machinas fallantes: a música gravada em Portugal no início do século XX* (Lisboa: Tinta da China, 2013), 17 and 21.
5. Benedetta Zucconi, *Coscienza fonografica. La riflessione sul suono registrato nell'Italia del primo Novecento* (Napoli-Salerno: Orthotes, 2018).
6. Mariano Gómez-Montejano, *El fonógrafo en España. Cilindros españoles* (Madrid: self-published, 2005).
7. Josep Perramon i Basqué, "Presentació de l'Aspe, Associació per a la Salvaguarda del Patrimoni Enregistrar," *Girant a 78 rpm*, 1:1 (2002), 3–4; Antoni Torrent i Marqués, "Girant al voltant dels Berliner's," *Girant a 78 rpm*, 1:3 (2003), 3–7.
8. Centro Andaluz del Flamenco, *Cilindros de cera, fondos del Centro Andaluz de Flamenco: primeras grabaciones del flamenco* (Mairena de Aljarafe: Calé Records, 2003); Javier Barreiro and Gabriel Marro, *Primeras grabaciones fonográficas en Aragón. 1898–1903. Una colección de cilindros de cera* (Zaragoza: Coda OUT/Gobierno de Aragón, 2007); Javier Barreiro, *Antiguas grabaciones fonográficas aragonesas, 1898–1907: la colección de cilindros para fonógrafo de Leandro Pérez* (Zaragoza: Coda OUT/Gobierno de Aragón, 2010); Sara Carrasco Mas, *Puesta en valor de la colección de cilindros de cera del Museo de Historia de la Telecomunicación Vicente Miralles Segarra* (undergraduate dissertation, Universitat Politècnica de València, 2018); Jaione Landaberea, "Ybarra Family Fonds," *Eresbil. Archivo Vasco de la Música*, http://www.eresbil.com/web/ybarra/Pagina.aspx?moduleID=2717&lang=en, accessed December 17, 2018.
9. Recent work includes: Samuel Llano, *Discordant Notes. Marginality and Social Control in Madrid, 1850–1930* (New York: Oxford University Press, 2018); Ian Biddle, "Madrid's Great Sonic Transformation: Sound, Noise, and the Auditory Commons of the City in the Nineteenth Century," *Journal of Spanish Cultural Studies*, 19:3 (2019), 227–40.
10. In the last few years there has indeed been evidence of performance studies and historical performance practice starting to attract interest in Spanish musicology,

particularly after the conference "Musicología aplicada al concierto: los estudios de *performance* en acción" (Musicology applied to the concert: Performance studies in action) took place in Baeza (Andalusia) in December 2016.

11. Cisneros Solá also refers to historical recordings of Manuel de Falla's art songs in María Dolores Cisneros Solá, "La obra para voz y piano de Manuel de Falla: Contexto artístico-cultural, proceso creativo y primera recepción" (Ph.D. diss., Universidad Complutense de Madrid, 2018).

12. V. K. Chew, *Talking Machines: 1877–1914: Some Aspects of the Early History of the Gramophone* (London: Her Majesty's Stationery Office, 1967); Oliver Read and Walter L. Welch, *From Tin Foil to Stereo Evolution of the Phonograph* (Indiana: Howard W. Sams, 1976); Roland Gelatt, *Edison's Fabulous Phonograph 1877–1977* (London: MacMillan, 1977); Daniel Marty, *Histoire illustré du phonographe* (Paris: Lazarus, 1979); Paul Charbon, *La machine parlante* (Paris: J. P. Gyss, 1981); Pekka Gronow, "The Record Industry: The Growth of a Mass Medium," *Popular Music*, 3 (1983), 53–75; James P. Kraft, *Stage to Studio. Musicians and the Sound Revolution, 1890–1950* (Baltimore and London: The Johns Hopkins University Press, 1996); David Patmore, "Selling Sounds: Recording and the Record Business," *The Cambridge Companion to Recorded Music*, ed. Nicholas Cook, Eric Clarke, Daniel Leech-Wilkinson, and John Rink (Cambridge: Cambridge University Press, 2009), 120–39; Andrew Blake, "Recording Practices and the Role of the Producer," *The Cambridge Companion to Recorded Music*, ed. Nicholas Cook, Eric Clarke, Daniel Leech-Wilkinson, and John Rink (Cambridge: Cambridge University Press, 2009), 36–53; David Suisman, *Selling Sounds. The Commercial Revolution in American Music* (Cambridge, MA: Harvard University Press, 2009); Martland, *Recording History*; Henri Chamoux, "La diffusion de l'enregistrement sonore en France à la Belle Époque (1893–1914), artistes, industriels et auditeurs du cylindre et du disque" (Ph.D. thesis, Université de Paris I, 2015); Susana Belchior, "Manufacturing Records," *The Lindstrom Project*, vol. 9, ed. Pekka Gronow, Christiane Hofer, and Mathias Böhm (Vienna: Gesellschaft für Historische Tonträger, 2017); Cañardo, *Fábricas de músicas*.

13. Michael Chanan, *Repeated Takes: A Short History of Recording and Its Effects on Music* (New York: Verso, 1995); Patrick Feaster, "Framing the Mechanical Voice: Generic Conventions of Early Sound Recording," *Folklore Forum*, 32 (2001), 57–102; Robert Philip, *Performing Music in the Age of Recording* (New Haven: Yale University Press, 2004); Daniel Leech-Wilkinson, "Portamento and Musical Meaning," *Journal of Musicological Research*, 25 (2006), 233–61; Nicholas Cook, "Performance Analysis and Chopin's Mazurkas," *Musicae Scientiae*, 11 (2007), 183–207; Daniel Leech-Wilkinson, "Sound and Meaning in Recordings of Schubert's 'Die junge Nonne'," *Musicae Scientiae*, 11 (2007), 209–36; Patrick Feaster, "'The Following Record': Making Sense of Phonographic Performance, 1877–1908" (Ph.D. diss., Indiana University, 2007); Rebecca Plack, "The Substance of Style: How Singing Creates Sound in Lieder Recordings, 1902–1939" (Ph.D. diss., Cornell University, 2008); Daniel Leech-Wilkinson, *The Changing Sound of Music* (London: CHARM, 2009); Nicholas Cook, "The Ghost in the Machine: Towards a Musicology of Recordings," *Musicae*

Scientiae, 14 (2010), 3–21; Nicholas Cook, *Beyond the Score: Music as Performance* (New York: Oxford University Press, 2013).

14. Thomas Y. Levin, "For the Record: Adorno on Music in the Age of Its Technological Reproducibility," *October*, 55 (1990), 23–47; Fred Moten, "The Phonographic *mise-en-scène*," *Cambridge Opera Journal*, 16 (2004), 269–81; Marc Perlman, "Golden Ears and Meter Readers: The Contest for Epistemic Authority in Audiophilia," *Social Studies of Science*, 34:5 (2004), 783–807; Bennett Hogg, "The Cultural Imagination of the Phonographic Voice, 1877–1940" (Ph.D. diss., University of Newcastle, 2008); Mark Katz, *Capturing Sound: How Technology Has Changed Music* (Berkeley: University of California Press, 2010); Arved Ashby, *Absolute Music, Mechanical Reproduction* (Oakland: University of California Press, 2010); Mark Katz, "The Amateur in the Age of Mechanical Music," *The Oxford Handbook of Sound Studies*, ed. Trevor Pinch and Karin Bijsterveld (New York: Oxford University Press, 2011), 459–78; Patrick Feaster, "'Rise and Obey the Command': Performative Fidelity and the Exercise of Phonographic Power," *Journal of Popular Music Studies*, 24 (2012), 357–95; João Silva, "Mechanical Instruments and Phonography: The Recording Angel of Historiography," *Radical Musicology*, 6 (2012–2013), http://www.radical-musicology.org.uk/2012/DaSilva.htm (accessed June 2020); Stefan Gauß, "Listening to the Horn: On the Cultural History of the Phonograph and the Gramophone," *Sounds of Modern History: Auditory Cultures in 19th- and 20th-Century Europe*, ed. Daniel Morat (Oxford, New York: Berghahn, 2014), 71–100; Michael Denning, *Noise Uprising. The Audiopolitics of a World Musical Revolution* (New York: Verso, 2015); Juan Fernando Velásquez Ospina, "(Re)sounding Cities: Urban Modernization, Listening, and Sounding Cultures in Colombia, 1886–1930" (Ph.D. diss., University of Pittsburgh, 2018); Sergio Ospina-Romero, "Ghosts in the Machine and Other Tales around a 'Marvelous Invention': Player Pianos in Latin America in the Early Twentieth Century," *Journal of the American Musicological Society*, 72:1 (2019), 1–42.

15. Lisa Gitelman, *Scripts, Grooves, and Writing Machines. Representing Technology in the Edison Era* (Stanford: Stanford University Press, 1999); William Howland Kenney, *Recorded Music in American Life: The Phonograph and Popular Memory, 1890–1945* (New York and Oxford: Oxford University Press, 1999); John M. Picker, "The Victorian Aura of the Recorded Voice," *New Literary History*, 32 (2001), 769–86; René T. A. Lysloff and Leslie C. Gay, Jr., eds., *Music and Technoculture* (Middletown, CT: Wesleyan University Press, 2003); Jonathan Sterne, *The Audible Past. Cultural Origins of Sound Reproduction* (Durham and London: Duke University Press, 2003); John M. Picker, *Victorian Soundscapes* (New York: Oxford University Press, 2003); Gustavus Stadler, "Never Heard Such a Thing. Lynching and Phonographic Modernity," *Social Text*, 28 (2010), 87–110; Trevor Pinch and Karin Bijsterveld, "New Keys to the World of Sound," *The Oxford Handbook of Sound Studies*, ed. Trevor Pinch and Karin Bijsterveld (New York: Oxford University Press, 2011), 3–36; Julia Kursell, "A Gray Box: The Phonograph in Laboratory Experiments and Fieldwork, 1900–1920," *The Oxford Handbook of Sound Studies*, ed. Trevor Pinch and Karin Bijsterveld (New York: Oxford University Press, 2011), 176–94.

16. Joseph Harrison, "Tackling National Decadence: Economic Regenerationism in Spain after the Colonial Debacle," *Spain's 1898 Crisis. Regenerationism, Modernism, Post-Colonialism*, ed. Joseph Harrison and Alan Hoyle (Manchester and New York: Manchester University Press, 2000), 55–67.

17. José Ferrater Mora, *Three Spanish Philosophers* (Albany: State University of New York Press, 2003), 64.

18. Ferrater Mora, *Three Spanish Philosophers*, 64.

19. José Andrés-Gallego, *Un 98 distinto. Regeneración, desastre, regeneracionismo* (Madrid: Ediciones Encuentro, 1998), 242.

20. Jesús Pando y Valle, *Regeneración económica. Croquis de un libro para el pueblo* (Madrid: Imprenta de Ricardo Rojas, 1897).

21. Antonio Royo Villanova, *La regeneración. El problema político* (Madrid: Hijos de M.G. Hernández, 1899).

22. Andrés-Gallego, *Un 98 distinto*, 241; Vicente Salavert and Manuel Suárez Cortina, "Introducción," *El regeneracionismo en España. Política, educación, ciencia y sociedad*, ed. Vicente Salavert and Manuel Suárez Cortina (Valencia: Universitat de Valencia, 2007), 9–19 (p. 11).

23. Andrés-Gallego, *Un 98 distinto*, 253; Manuel Suárez Cortina, "Sociedad, cultura y política en la España de entre siglos," *El regeneracionismo en España. Política, educación, ciencia y sociedad*, ed. Vicente Salavert and Manuel Suárez Cortina (Valencia: Universitat de Valencia, 2007), 21–46 (p. 22).

24. José Manuel Sánchez Ron, "Investigación científica, desarrollo tecnológico y educación en España (1900–1950)," *Arbor*, 96 (1992), 33–74 (p. 35).

25. Alfredo Baratas Díaz, "La ciencia española ante la crisis del 98: semillas, frutos y agostamiento," *Cuadernos de Historia Contemporánea*, 20 (1998), 151–63.

26. José María López Piñero, ed., *La ciencia en la España del siglo XIX* (Madrid: Marcial Pons, 1992).

27. Santos Casado de Otaola, *Naturaleza patria: ciencia y sentimiento de la naturaleza en la España del Regeneracionismo* (Madrid: Marcial Pons, 2010), 178.

28. Casado de Otaola, *Naturaleza patria*, 97.

29. Casado de Otaola, *Naturaleza patria*, 272.

30. Karolina Doughy and Maja Lagerkvist, "The Pan Flute Musicians at Sergels Torg: Between Global Flows and Specificities of Place," *Researching and Representing Mobilities. Transdisciplinary Encounters*, ed. Lesley Murray and Sara Upstone (London: Palgrave MacMillan, 2014), 150–69 (pp. 150–1).

31. Sumanth Gopinath and Jason Stanyek, "Anytime, Anywhere? An Introduction to the Devices, Markets, and Theories of Mobile Music," *The Oxford Handbook of Mobile Music Studies*, vol. 1, ed. Sumanth Gopinath and Jason Stanyek (New York: Oxford University Press, 2014), 1–34 (p. 25).

32. Sánchez Ron, "Investigación científica, desarrollo tecnológico y educación en España," 38.

33. Bernardo Riego Amezaga, "Imágenes fotográficas y estrategias de opinión pública: los viajes de la Reina Isabel II por España (1858–1866)," *Patrimonio Nacional*, 139 (1999), 2–13.

34. Lily Litvak, *El tiempo de los trenes: el paisaje español en el arte y la literatura del realismo: (1849–1918)* (Madrid: Ediciones del Serbal, 1991), 12.
35. Litvak, *El tiempo de los trenes*, 184.
36. Alberto Elena and Javier Ordóñez, "Science, Technology, and the Spanish Colonial Experience in the Nineteenth Century," *Osiris*, 15 (2000), 70–82.
37. Litvak, *El tiempo de los trenes*, 16 and 181, 194–5, 213.
38. "The Talking Phonograph," *Scientific American*, December 22, 1877, 300.
39. Gelatt, *Edison's Fabulous Phonograph*, 27.
40. Chew, *Talking Machines*, 13–14.
41. Gronow, "The Record Industry," 56.

Chapter 1

1. Literally, "repetition telephone." This was the name given to the phonograph in early news articles in Spain.
2. José Alcover, "El teléfono Bell y el teléfono de repetición de M. Edison," *La gaceta industrial*, December 10, 1877, 2–4. Alcover's source was Edward H. Johnson, "A Wonderful Invention. Speech Capable of Infinite Repetition from Automatic Records," *Scientific American*, November 17, 1877, 304. The *Scientific American* article was translated (presumably by Alcover himself) and quoted in its entirety.
3. Read and Welch, *From Tin Foil to Stereo*, 26.
4. Read and Welch, *From Tin Foil to Stereo*, 26.
5. "El fonógrafo," *El constitucional*, January 23, 1878, 1; "Sección de noticias," *El progreso de Lugo*, January 26, 1878, 3.
6. Mercedes Fernández Paradas, "La industria eléctrica y su actividad en el negocio del alumbrado en España (1901–1935)," *Ayer*, 71 (2008), 245–65 (p. 246).
7. Pedro Ruiz-Castell, "Scientific Instruments for Education in Early Twentieth-Century Spain," *Annals of Science*, 65:4 (2008), 519–27 (p. 522).
8. Antoni Torrent i Marqués, "Els primers enregistraments a casa nostra," *Girant a 78 rpm*, 11 (2008), 6–13 (p. 6); "Ateneo Barcelonés," *La publicidad*, September 9, 1878, 1.
9. *Diari catalá*, July 24, 1879, 3; *Diari catalá*, June 25, 1879, 3.
10. *Diari catalá*, September 21, 1879, 3.
11. *Diari catalá*, September 28, 1880, 3.
12. "Popularisasió del teléfono," *Diari catalá*, September 24, 1880, 3.
13. *Lo gay saber*, November 1, 1880, 1.
14. Record collector Torrent i Marqués named Dalmau "the first importer of phonographs in Catalonia" (and, by extension, all of Spain); see Torrent i Marqués, "Els primers enregistraments a casa nostra", 6. The term "importer" (*importador*) seems to suggest that Dalmau was importing significant numbers of devices for commercial purposes, but it does not seem plausible that Dalmau had been engaged in such activities at the time when he was conducting demonstrations, as the phonograph was seen at the time as having limited commercial appeal for sale among the general public. By

1886 there was at least one importer of phonographs in Europe—a British instrument dealer who advertised three different models of phonographs (Chew, *Talking Machines: 1877–1914*, 6), but there is no evidence that Dalmau or others in Spain were importing phonographs for sale at around that time.

15. "Revista de la semana," *La mañana*, January 19, 1879, 1.
16. "Noticias generales," *La época*, January 12, 1879, 4.
17. Sociedad El Fonógrafo, *Sesiones de fonógrafo* (advertisement), *La correspondencia de España*, January 31, 1879, 1.
18. "Edición de la noche," *La correspondencia de España*, February 6, 1879, 2.
19. "Semana histórica," *La academia*, February 15, 1879, 3.
20. The last advertisement published in the Madrid press was Sociedad El Fonógrafo, *Sesiones de fonógrafo* (advertisement), *El campo*, February 16, 1879, 16.
21. Even though Bargeon de Viverols's demonstrations and presumably interests differed from those of Dalmau's in a number of ways, there is no evidence of friction between both men, or that they were even aware of each other (although it is likely that Dalmau at least knew about the Frenchman, given that Bargeon de Viverols visited Barcelona).
22. *La Publicidad*, April 4, 1879, 3; "Cridas," *Lo nunci*, April 20, 1879, 4.
23. V. P. T., "El fonógrafo," *Teléfono catalán*, May 25, 1879, 1.
24. "Noticias locales y generales," *El comercio*, June 10, 1879, 2; "Boletín de teatros," *La correspondencia de España*, June 23, 1879, 1; Jean-Marie Bargeon de Viverols, *Teatro de Apolo. Funciones para hoy martes 24 de junio de 1879* (advertisement) (June 24, 1879), http://bdh-rd.bne.es/viewer.vm?id=0000032906 (accessed May 21, 2018).
25. *El Papa-Moscas*, 74 (1879), 2.
26. Jean-Marie Bargeon de Viverols, *Única representación del fonógrafo Edison. El invento más admirable del siglo* (advertisement) (1879), clip from an unknown publication held at the Biblioteca de Catalunya, reference number Cat 1/7.
27. "Sesiones," *Diario de Córdoba*, January 20, 1880, 1.
28. "Fonógrafo Edison," *Diario de Córdoba*, February 12, 1880, 2; "Gacetillas," *El graduador*, February 19, 1880, 2; "Crónica," *El eco de Cartagena*, February 28, 1880, 2.
29. "Liceo," *Diario de Murcia*, March 3, 1880, 1.
30. "Gacetillas," *Crónica de Cataluña*, April 7, 1886, 1. It is likely that, in the same way as Dalmau, Bargeon de Viverols thought, after his initial experiments with phonographs, that the device's potentialities would remain limited until the technology was developed further.
31. Solange Hibbs and Caroline Fillière, "Introducción," *Los discursos de la ciencia y de la literatura en España (1875–1906)*, ed. Solange Hibbs and Caroline Fillière (Madrid: Academia del Hispanismo, 2015), 13–28 (p. 15).
32. Catherine Sablonnière, "Literatura científica y vulgarización de las ciencias ante la crisis de la ciencia en la España finisecular: ¿hacia una poética del discurso científico?," *Los discursos de la ciencia y de la literatura en España (1875–1906)*, ed. Solange Hibbs and Caroline Fillière (Madrid: Academia del Hispanismo, 2015), 305–18 (pp. 313–14). See also, on music hall and mesmerism in the British context David Knight, "La

popularización de la ciencia en la Inglaterra del siglo XIX," *La ciencia y su público*, ed. Javier Ordóñez y Alberto Elena (Madrid: CSIC, 1990), 311–30 (p. 323).

33. Fernández, *Tecnología, espectáculo y literatura*, 65; Robin E. Rider, "El experimento como espectáculo," *La ciencia y su público*, ed. Javier Ordóñez y Alberto Elena (Madrid: CSIC, 1990), 113–45 (pp. 116–17).

34. Carl Willmann, *La magia moderna de salón* (Valencia: Pascual Eguilar, 1897), 1.

35. Willmann, *La magia moderna de salón*, 47.

36. Willmann, *La magia moderna de salón*, 39.

37. Allan Kardec, *Espiritismo experimental: El libro de los médiums. Guía de los médiums y de las evocaciones* (Barcelona: Carbonell y Esteva, 1904), 19–25.

38. Kardec, *Espiritismo experimental*, 35.

39. Armando Baeza y Salvador, *Ciencia popular: los misterios de la ciencia* (Barcelona: Ramón Molinas, 1890).

40. These included *cosmorama, mundonuevo, panorama, diorama, linterna mágica, monstruos gigantescos*, and *cuadros disolventes*. Luis Miguel Fernández, *Tecnología, espectáculo, literatura: dispositivos ópticos en las letras españolas de los siglos XVIII y XIX* (Santiago de Compostela: Publicacións Universidade de Santiago de Compostela, 2006), 105–318.

41. Bernardo Riego Amezaga, "Visibilidades diferenciadas: usos sociales de las imágenes en la España isabelina," *Ojos que ven, ojos que leen. Textos e imágenes en la España Isabelina*, ed. Marie-Linda Ortega (Madrid: Visor/Presses Universitaires de Marne-La-Vallée, 2004), 57–66 (pp. 57–60 and 64).

42. Bargeon de Viverols, *Teatro de Apolo. Funciones para hoy martes 24 de junio de 1879* (advertisement).

43. Gelatt, *Edison's fabulous phonograph*, 27.

44. Jean-Marie Bargeon de Viverols, telegram to Thomas Alva Edison, October 11, 1888. The Thomas Edison Papers, Rutgers University, D8849, http://edison.rutgers.edu/NamesSearch/DocImage.php?DocId=D8849ACK& (last accessed May 22, 2018).

45. Bargeon de Viverols, *Teatro de Apolo. Funciones para hoy martes 24 de junio de 1879* (advertisement).

46. It is not clear what this refers to; Bargeon de Viverols's other publicity or reviews do not shed any light on this. It could refer to a recording of a person reciting a series of arithmetic operations, which were then played back by the phonograph. Nevertheless, whereas all of the other numbers advertised by Bargeon de Viverols (comic songs, brass solos, military commands, etc.) were popular too in sessions throughout the 1890s, I have not been able to find any records of anything comparable to this "arithmetic of the phonograph" (*aritmética del fonógrafo*).

47. Hibbs and Fillière, "Introducción", 15; Borja Rodríguez Gutiérrez, "Ciencia e imágenes de la ciencia," *Los discursos de la ciencia y de la literatura en España (1875–1906)*, ed. Solange Hibbs y Caroline Fillière (Madrid: Academia del Hispanismo, 2015), 225–46; Paula Carabell, "Photography, Phonography and the Lost Object," *Perspectives of New Music*, 40:1 (2002), 176–89 (pp. 181 and 187).

48. "Conferencias en Círculo de la Unión Católica," *El magisterio español*, January 20, 1882, 2.

49. "Jochs florals," *La ilustració catalana*, February 1, 1882, 13. The same prize of a phonograph was offered the next year: "Jochs florals," *La ilustració catalana*, January 30, 1883, 14.

50. *La correspondencia de España*, February 10, 1882, 2, "El fonógrafo", *El día*, August 19, 1882, 3.

51. José Estremera, "Microscopio gigante," *Madrid cómico*, August 31, 1884, 2; *El liberal*, April 17, 1885, 3; "Jardín del buen Retiro," *El popular*, August 22, 1884, 3; "Sección de espectáculos," *El Busilis*, December 12, 1884, 4.

52. "Teatro de la Comedia: exhibición de fonógrafo por el Dr. Llop [sic]," *El globo*, February 14, 1882, 3; "Diversiones públicas," *El liberal*, February 14, 1882, 3.

53. "El fonógrafo," *El día*, August 19, 1882, 3.

54. *La correspondencia de España*, August 19, 1882, 3.

55. *La discusión*, August 20, 1882, 3; *Crónica de la música*, August 23, 1882, 1.

56. "El fonógrafo," *El día*, August 19, 1882, 3.

57. *El liberal*, August 29, 1882, 5; *El imparcial*, September 14, 1882, 2; [First name unknown] López Alué, "De La Granja," *El imparcial*, September 18, 1882, 2; "Edición de la tarde," *La correspondencia de España*, September 20, 1882, 2.

58. The first advertisement I have been able to locate is from *El eco nacional*, May 8, 1885, 3, with the last one being in *La correspondencia de España*, August 31, 1885, 4.

59. *El nuevo Ateneo*, March 15, 1887, 5.

60. Pinch and Bijsterveld, "New Keys to the World of Sound," 5.

61. Eric W. Rothenbuhler and John Durham Peters, "Defining Phonography: An Experiment in Theory," *The Musical Quarterly*, 81:2 (1997), 242–64 (p. 245).

62. Gauß, "Listening to the Horn," 77; Jacques Vest, "Vox Machinae: Phonographs and the Birth of Sonic Modernity, 1877–1930" (Ph.D. diss., University of Michigan, 2018), 9.

63. Joaquín Martín de Sagarmínaga, *Diccionario de cantantes líricos españoles* (Madrid: Fundación Caja de Madrid, 1997), 155.

64. Mariano Gómez-Montejano, "El fonógrafo en España (segunda parte)," *Girant a 78 rpm*, 6 (2004), 9–11 (p. 9).

65. Picker, *Victorian Soundscapes*, 111; Picker, "The Victorian Aura of the Recorded Voice," 769.

66. Un diletante [pseudonym], "Música clásica. Cuartetos en el Conservatorio," *La correspondencia de España*, December 16, 1880, 2 and 4.

67. "Miscelánea," *El Ateneo*, March 21, 1878, 8; "Semana histórica," *La Academia*, March 23, 1878, 2.

68. José Casas y Barbosa, "Descripción del teléfono, el micrófono y el fonógrafo," *Las maravillas de la naturaleza, de la ciencia y del arte*, 5th ed., ed. Manuel Aranda y Sanjuán (Barcelona: Trilla y Serrra, no year). The fifth edition is not dated, and the catalogue of the Biblioteca Nacional de España gives it as 1875; however, given that it includes a description of the phonograph, it must by force date from after 1877.

69. Casas y Barbosa, "Descripción del teléfono, el micrófono y el fonógrafo," 66.

70. Gitelman, *Scripts, Grooves, and Writing Machines*, 2.

71. Emily Thompson, "Machines, Music, and the Quest for Fidelity: Marketing the Edison Phonograph in America, 1877–1925," *The Musical Quarterly*, 79:1 (1995), 131–71.

72. Bernardo Riego Amezaga, *La construcción social de la realidad a través de la fotografía y el grabado informativo en la España del siglo XIX* (Santander: Universidad de Cantabria, 2001), 22.

73. Bernardo Riego Amezaga, "La representación en las imágenes fotográficas: discursos y divergencias de cada tiempo histórico," *Fotografía e Historia. III Encuentro en Castilla la Mancha*, ed. Irma Fuencisla Álvarez Delgado and Ángel Luis López Villaverde (Cuenca: UCLM, 2009), 67–81 (p. 68); Riego Amezaga, *La construcción social*, 303.

74. "Gacetillas," *El graduador*, February 24, 1880, 2; "Teatro de la Comedia: exhibición de fonógrafo por el Dr. Llop [sic]," *El globo*, February 14, 1882, 3. The former article refers to an exhibition by Bargeon de Viverols, whereas the latter is about Dr. Llops.

75. A rare instance of a reviewer doubting the capabilities of the phonograph can be found (about one of Dalmau's demonstrations) in *Lo gay saber*, November 10, 1880, 11.

76. Pascual del Solorno, "El fonógrafo," *Revista del Ateneo Escolar de Guadalajara*, August 5, 1882, 53–5.

77. A. G. M., "El fonógrafo," *La semana católica de Salamanca*, June 5, 1886, 10.

78. Douglas Kahn, "Introduction," *Wireless Imagination. Sound, Radio, and the Avant-Garde*, ed. Douglas Kahn and Gregory Whitehead (Cambridge, MA, and London: The MIT press, 1992), 1–30 (p. 6).

79. Fernández, *Tecnología, espectáculo y literatura*, 27 and 49–65.

80. Josep Verdú, "La cansó del fonógrafo," *La Renaxensa*, October 30, 1879, 38–40.

81. "Descargas cerradas," *El fígaro*, February 14, 1880, 1; *El constitucional*, December 16, 1887, 1.

82. "Novedades teatrales," *El globo*, January 4, 1880, 3–4.

83. *La correspondencia de España*, June 20, 1885, 2; *La unión*, August 29, 1885, 3; "En Recoletos: *El fonógrafo*, obra de teatro de Castillo y Soriano," *El Imparcial*, August 30, 1885, 3; "El fonógrafo," *La correspondencia de España*, September 9, 1885, 2.

84. *La Iberia*, August 31, 1885, 3.

85. Publio Heredia y Larrea, *El testamento fonográfico* (Madrid: La Revista Política, 1895). See also X., "Sección varia. El fonógrafo," *La antorcha*, February 5, 1883, 3, which also advocated for introducing phonographic wills.

86. Id., p. 21.

87. Id., p. 22.

88. Id., p. 31.

89. Id., p. 114.

90. Id., p. 117.

Chapter 2

1. Thomas A. Edison, "The Perfected Phonograph," *The North American Review*, 146:379 (1888), 641–50.
2. Thomas A. Edison, "El nuevo fonógrafo de Edisson [sic]," *La publicidad*, August 2, 1888, 2. The name of the translator is not given. The phrase "fonógrafo perfeccionado" was used in Spanish newspapers before and after the invention of the Perfected Phonograph to refer generally to any kind of improvements or modifications Edison had made to the phonograph (see for example R[first name unknown] de Sorrarain, "La ciencia actual," *La ilustración española e hispanoamericana*, August 1, 1888, 3). Nevertheless, the translation in *La publicidad* is the first unequivocal instance where "fonógrafo perfeccionado" is referring to the Perfected Phonograph.
3. "Provincias," *La justicia*, August 23, 1888, 3.
4. Thomas A. Edison, "El fonógrafo perfeccionado," *Crónica científica*, September 25, 1888, 411–14.
5. "El fonógrafo y el grafófono," *La gaceta industrial*, December 25, 1888, 9.
6. *Exposición Universal de Barcelona: año 1888. Clasificación de productos* (Barcelona: publisher unknown, 1887); *Exposición universal de Barcelona. Catálogo oficial especial de España* (Barcelona: Imprenta de los Sucesores de N. Ramírez, 1888); Manuel Girona, *Memoria Reglamentaria de la Exposición Universal de Barcelona de 1888* (Barcelona: Heinrich, 1889). Gómez-Montejano also argues that the phonograph was not exhibited at the Exposición. See Gómez-Montejano, "El fonógrafo en España (segunda parte)," 9.
7. "Una feria fantástica," *Diario de Tenerife*, January 29, 1890, 4.
8. "Espectáculos," *La publicidad*, February 2, 1890, 3; "Teatro Circo," *La vanguardia*, February 28, 1890, 2; "Espectáculos," *La publicidad*, March 2, 1890, 3.
9. "Noticias y avisos," *La nueva lucha: diario de Gerona*, March 4, 1890, 3.
10. "Noticias," *La correspondencia de España*, November 28, 1897, 3; "Fonógrafo y cinematógrafo," *La vanguardia*, March 4, 1898, 2.
11. Catalogued under Biblioteca Nacional de Catalunya, *Cilindres de cera de la col·lecció Mariano Gómez Montejano* [non-commercial CD], Biblioteca Nacional de Catalunya, CD 4175.
12. "Crónica general," *Crónica reusense*, July 12, 1897, 1.
13. http://hemerotecadigital.bne.es, http://prensahistorica.mcu.es, http://mdc1.csuc.cat.
14. "De Ciudadela," *El grano de arena*, June 7, 1905, 3.
15. "Eliseo Exprés," *La correspondencia alicantina*, May 7, 1897, 2.
16. Gelatt, *Edison's Fabulous Phonograph*, 27; Read and Welch, *From Tin Foil to Stereo*, 49–50 and 54.
17. Questel—Edison workshop in Paris, *Fonógrafo* (advertisement), *El Imparcial*, July 2, 1895, 4.
18. "Fonógrafo," *El Cantábrico*, August 29, 1896, 3.
19. "Gacetilla," *El bien público*, June 28, 1892, 2; "Gacetilla," *El liberal*, July 2, 1892, 3.
20. "Noticias," *La correspondencia de España*, December 1, 1892, 3.
21. "Gacetillas," *El eco de Navarra*, July 7, 1895, 2.

22. "Espectáculos," *La vanguardia*, September 29, 1896, 7; "Teatro Principal," *El graduador*, May 18, 1897, 3; "Noticias locales," *El regional*, July 28, 1897, 2.

23. "Sección de noticias," *La Rioja*, June 17, 1894, 2; "Sección de noticias," *La Rioja*, September 19, 1894, 2.

24. "Ecos de Sóller," *Heraldo de Baleares*, August 5, 1895, 2; "Noticias," *La lucha*, January 14, 1896, 3; "Noticias," *El baluarte*, February 23, 1896, 3; "Gacetillas," *El vigía católico*, June 27, 1896, 3.

25. "Gacetillas," *El eco de Navarra*, November 20, 1896, 2.

26. "Leemos en *La Monarquía*," *El nuevo alicantino*, July 8, 1896, 3; "Un fonógrafo," *El Ateneo*, July 10, 1896, 7; "Fonógrafo Edison," *El graduador*, July 20, 1898, 20.

27. "Fonda de Oriente," *Diario de Córdoba*, March 21, 1894, 3; "Crónica local," *Heraldo de Baleares*, October 19, 1895, 2; "Ateneo de Madrid," *El globo*, December 16, 1895, 3; "Gacetillas," *Crónica meridional*, January 31, 1896, 3; "Gacetillas," *El Guadalete*, March 7, 1896, 2; "Gacetillas," *El bien público*, October 7, 1896, 3.

28. "Pórtico de Apolo," *La correspondencia de España*, October 19, 1896, 3.

29. A. [first name unknown] Mathieu-Villon, *El fonógrafo* (Madrid: A. Avial, 1893).

30. Fonógrafo Soley, *Fonógrafo Soley* (advertisement), *La publicidad*, August 30, 1893, 7.

31. "Gacetillas," *El heraldo de Madrid*, January 11, 1894, 4.

32. "Espectáculos," April 17, 1894, 4.

33. "Madrid," *La correspondencia de España*, April 28, 1895, 2.

34. "Avisos y noticias," *La Rioja*, October 6, 1897, 2; "Salón Expréss. Pórticos de Berdejo," *La voz de la provincia*, November 27, 1897, 3; "Noticias," *Heraldo de Navarra*, July 13, 1898; "Gacetillas," *El eco de Navarra*, August 5, 1898, 2; "Noticias," *La Rioja*, September 25, 1898, 2; "Salón Exprés," *El noticiero de Soria*, October 29, 1898, 2; "Salón Exprés," *El noticiero de Soria*, November 2, 1898, 2.

35. "Accésit," *El noticiero de Soria*, October 3, 1896, 2.

36. K., "Crónicas madrileñas," *La correspondencia de España*, March 12, 1897, 1; "Desde la corte," *El isleño*, August 2, 1897, 1; Barón de Stoff, "Noticias telegráficas," *El globo*, February 3, 1897, 2; "Exposición artística a beneficio de los heridos de Cuba y Filipinas," *La época*, February 24, 1897, 3.

37. *El bien público*, June 28, 1892.

38. "Anuncios," *El eco de Santiago*, October 28, 1898, 2.

39. "Edición de la tarde," *Heraldo de Baleares*, April 2, 1895, 3; "Menestra," *La Rioja*, April 19, 1896, 2.

40. "Noticias," *La correspondencia de Alicante*, May 11, 1897, 6.

41. "Sucesos," *La lealtad navarra*, July 6, 1893, 2.

42. "Noticias," *La atalaya*, January 28, 1894, 3; "Fonógrafo," *Diario de Córdoba*, August 20, 1892, 3.

43. "Crónica general," *La dinastía*, September 7, 1895, 2; "Crónica general," *El adelanto de Salamanca*, August 24, 1903, 1; Pitis, "Un charro filósofo," *El adelanto de Salamanca*, August 24, 1903, 1; Antonio de la Cuesta y Sáinz, "Cuento," *El correo ibérico*, December 30, 1904, 2.

44. Miguel Martorell and Santos Juliá, *Manual de Historia Política y Social de España (1808–2018)* (Barcelona: RBA Libros, 2019), 92.

45. R. Blasco [name unknown], "*El fonógrafo ambulante*," *La correspondencia de España*, April 25, 1899, 3.

46. "Crónica de espectáculos," *El día de Palencia*, May 11, 1899, 3.

47. As has been discussed previously, the Marquis of Tovar was another aristocrat with an early interest in the phonograph. He, however, died eight months before *El fonógrafo ambulante* was premiered, which makes him a more remote possibility (particularly considering that *El día de Palencia* claimed that the librettist was indeed in the theater during the premiere).

48. I examine the *zarzuela* in more detail, expanding upon the preceding arguments, in the following article: Eva Moreda Rodríguez, "Travelling Phonographs in *fin de siècle* Spain: Recording Technologies and National Regeneration in Ruperto Chapí's *El fonógrafo ambulante*," *Journal of Spanish Cultural Studies*, 20:3 (2019), 241–55.

49. Moreda Rodríguez, "Travelling Phonographs in *fin de siècle* Spain," 250–1.

50. Moreda Rodríguez, "Travelling Phonographs in *fin de siècle* Spain," 243–4.

51. Álvaro Girón Sierra and Jorge Molero-Mesa, "The Rose of Fire. Anarchist Culture, Urban Spaces and Management of Scientific Knowledge in a Divided City," *Barcelona: An Urban History of Science and Modernity, 1888–1929*, ed. Oliver Hochadel and Agustí Nieto-Catalán (Oxford and New York: Routledge, 2006), 115–35 (p. 119); Oliver Hochadel and Agustí Nieto-Catalán, "Introduction," *Barcelona: An Urban History of Science and Modernity, 1888–1929*, ed. Oliver Hochadel and Agustí Nieto-Catalán (Oxford and New York: Routledge, 2006), 1–21 (p. 13).

52. "Círculo militar," *La correspondencia de España*, November 19, 1899, 3; "Noticias de provincias," *El globo*, April 30, 1905, 2.

53. Such as Francisco Crespo Hidalgo, who exhibited a phonograph in some localities in Galicia in 1900; see "Audiciones fonográficas," *El país*, April 17, 1900, 4. *Gabinete* owners with an army background include Álvaro Ureña and Atanasio Palacio Valdés, who will be both briefly discussed in Chapter 3.

54. "Notas locales," *La vanguardia*, March 5, 1890, 7.

55. Jean-Louis Guereña, "La sociabilidad en la España contemporánea," *Sociabilidad fin de siglo. Espacios asociativos en torno a 1898*, ed. Isidro Sánchez Sánchez and Rafael Villena Espinosa (Cuenca: Ediciones de la Universidad de Castilla-La Mancha, 1999), 15–44 (p. 26); Pere Solà, *Els Ateneus obrers i la cultura popular a Catalunya (1900–1939): l'Ateneu Enciclopèdic Popular* (Barcelona: La Magrana, 1978), 16, 38, and 46.

56. Guereña, "La sociabilidad en la España contemporánea," 29; Mari Carmen Naranjo Santana, *Cultura, ciencia y sociabilidad en Las Palmas de Gran Canaria en el siglo XIX: el Gabinete Literario y el Museo Canario* (Rivas Vaciamadrid: Mercurio, 2016), 55.

57. José Gregorio Cayuela Fernández, "Proyectos de sociedad y nación: la crisis del concepto de España en el 98," *Sociabilidad fin de siglo. Espacios asociativos en torno a 1898*, ed. Isidro Sánchez Sánchez and Rafael Villena Espinosa (Cuenca: Ediciones de la Universidad de Castilla-La Mancha, 1999), 45–72.

58. Isidro Sánchez Sánchez, "Las luces del 98. Sociedades eléctricas en la España finisecular," *Sociabilidad fin de siglo. Espacios asociativos en torno a 1898*, ed. Isidro

Sánchez Sánchez and Rafael Villena Espinosa (Cuenca: Ediciones de la Universidad de Castilla-La Mancha, 1999), 151–223.

59. María Zozaya, *Identidades en juego: formas de representación social del poder de la élite en un espacio de sociabilidad masculino, 1836–1936* (Madrid: Siglo XXI de España, 2015), 15; Rafael Serrano García, *El Círculo de Recreo de Valladolid (1844–2010): ocio y sociabilidad en un espacio exclusivo* (Valladolid: Universidad de Valladolid, 2011), 20.

60. Guereña, "La sociabilidad en la España contemporánea," 36.

61. Solà, *Els Ateneus obrers*, 46; Juan Uña y Sarthou, *Las asociaciones obreras en España* (Madrid: G. Juste, 1900), 333; Naranjo Santana, *Cultura, ciencia y sociabilidad en Las Palmas de Gran Canaria*, 24. On Catholic associations more specifically and their liberal-bourgeois shift in some contexts: Julio de la Cueva Merino, "Clericalismo y movilización católica durante la restauración," *Clericalismo y asociacionismo católico en España: de la restauración a la transición: un siglo entre el palio y el consiliario*, ed. Julio de la Cueva Merino and Ángel Luis López Villaverde (Cuenca: Ediciones de la Universidad de Castilla la Mancha, 2005), 27–50 (p. 31).

62. Enrique Mhartín y Guix, *Reglamentación oficial de sociedades, reuniones y manifestaciones públicas* (Barcelona: José Cunill y Sala, 1902), 8–10.

63. Zozaya, *Identidades en juego*, 57.

64. Naranjo Santana, *Cultura, ciencia y sociabilidad*, 45–6. There is no evidence that any casino permanently installed a phonograph in its premises during these years; nevertheless, the Madrid casino did experiment with the so-called teatrófono—a phone which allowed members to listen to operas performed at the Teatro Real (Zozaya, *Identidades en juego*, 173–4).

65. "Gacetillas," *El Guadalete*, March 5, 1895, 3.

66. "Audiencia provincial," *Flores y abejas*, May 22, 1898, 6.

67. *La nueva lucha*, March 4, 1890.

68. "Gacetillas," *El Guadalete*, March 16, 1895, 2.

69. "La velada del Círculo Mercantil," *La correspondencia de España*, December 19, 1897, 1.

70. "Noticias," *La opinión*, September 22, 1897, 2.

71. "Noticias," *El guasón*, January 12, 1896, 3; "Provincias," *El correo de España*, January 5, 1896, 2.

72. "Ecos madrileños," *La época*, July 6, 1899, 2; "Gacetilla," *El vigía de Ciudadela*, July 22, 1899, 2.

73. "El Ateneo," *El País*, December 15, 1895, 3; "Madrid," *El Globo*, December 16, 1895, 3.

74. "Teatro de San Andrés de Palomar," *La vanguardia*, September 6, 1890, 2.

75. "Provincias," *La correspondencia de España*, December 14, 1896, 4.

76. "Fonógrafo," *La semana católica de Salamanca*, December 5, 1896, 15.

77. "Círculos y sociedades," *Las provincias de Valencia*, February 22, 1899, 3.

78. *El Globo*, April 30, 1905.

79. Serrano García, *El Círculo de Recreo de Valladolid*, 45.

80. "El fonógrafo Edison," *Diario de Córdoba*, March 30, 1894, 3.

81. "El fonógrafo testigo," *La lucha*, November 8, 1895, 3.

82. *La Rioja*, June 24, 1894.

83. *La Rioja*, June 24, 1894.

84. See for example: "En el Ateneo," *Diario del comercio*, March 14, 1897, 2.

85. "Crónica general," *Crónica reusense*, July 10, 1897, 1.

86. "Salón Exprés," *Heraldo de Alcoy*, March 11, 1900, 3. The entry price was 0.25 *pesetas*.

87. "Gran subasta," *El avisador numantino*, January 4, 1900, 4. This was indeed one of the cheapest recorded salons, at 5 *céntimos* per single audition. The phonograph that Hugens installed at Teatro Apolo charged 10 *céntimos* per single audition; see *La correspondencia de España*, October 19, 1896.

88. For example, in Orihuela one audition was sold at 10 *céntimos*, with seven for 50 *céntimos*: "Fonógrafo Edisson [sic]," *El Thader*, November 14, 1895, 2. Or at the Café Colón in Barcelona, one cylinder for 10 *céntimos* and three cylinders for 25 *céntimos*: "Café Colón," *La publicidad*, January 27, 1897, 4.

89. "Fonógrafo," *La Atalaya*, January 6, 1896, 3; "Asilo de Santa Cristina," *Diario oficial de avisos de Madrid*, July 4, 1897, 3; "Plaza del Teatro," *El liberal*, November 9, 1899, 3; "La Feria," *El día de Palencia*, May 20, 1899, 2.

90. Carmen del Moral Ruiz, *El género chico* (Madrid: Alianza, 2004), 54.

91. Del Moral Ruiz, *El género chico*, 55.

92. Del Moral Ruiz, *El género chico*, 55–6.

93. "Salón del Heraldo," *Las Baleares*, March 16, 1895, 3.

94. Teatro Circo Balear, *Teatro Circo Balear* (advertisement), *Las Baleares*, March 16, 1895, 3.

95. "La Rioja en Haro," *La Rioja*, September 5, 1894, 1.

96. "Noticias," *La correspondencia de Alicante*, May 4, 1898, 2.

97. "Fonógrafo," *La región extremeña*, October 23, 1894, 2.

98. "Gacetillas," *El eco de Navarra*, July 18, 1897, 2.

99. "Universal Exprés," *La vanguardia*, February 14, 1897, 7.

100. A. T., "Fonógrafos automáticos," *Boletín fonográfico*, 14 (1900), 216–17.

101. Thompson, "Machines, Music, and the Quest for Fidelity," 131–2; Kerim Yasar, *Electrified Voices. How the Telephone, Phonograph and Radio Shaped Modern Japan, 1898–1945* (New York and Chichester: Columbia University Press, 2018).

102. "Noticias," *El carpetano*, July 2, 1896, 4; "Crónica," *El liberal navarro*, July 17, 1896, 2; "Gacetillas," *Diario de Córdoba*, February 2, 1898, 2.

103. "El fonógrafo perfeccionado de Edisson [sic]," *La correspondencia de España*, November 2, 1892, 3.

104. "Diversiones públicas," *La época*, November 4, 1892, 3.

105. "El fonógrafo en el Teatro," *La Rioja*, June 24, 1894, 2.

106. *El correo de España*, January 5, 1896.

107. *El bien público*, June 28, 1892; "Crónica," *El tradicionalista*, July 6, 1893, 6.

108. *La Rioja*, September 5, 1894.

109. *La correspondencia de España*, November 2, 1892.

110. "El fonógrafo," *Crónica meridional*, December 22, 1894, 2.

111. *La Rioja*, September 5, 1894.

112. "Notable fonógrafo," *El popular*, May 23, 1894, 2; "Noticias," *La unión católica*, May 25, 1894, 2; "Espectáculo científico," *La correspondencia de España*, June 27, 1894, 4;

"Fonógrafo," *La Iberia*, January 10, 1895, 3; "Noticias," *La correspondencia de España*, February 3, 1895, 2; "Madrid," *El correo militar*, February 11, 1895, 2; "Noticias," *El correo militar*, March 9, 1895, 2; "Noticias," *El día*, March 18, 1895, 3.

113. "El kinetóscopo de Edisson [sic]," *El día*, May 14, 1895, 3; "El fonógrafo de Segovia," *La legalidad*, December 7, 1895, 3; "Fonógrafo," *La tempestad*, September 6, 1896, 3.

114. "Crónica general," *La publicidad*, January 8, 1895, 2.

115. "Noticias de espectáculos," *La unión católica*, July 8, 1895, 3.

116. *La vanguardia*, March 4, 1898.

117. Colís is also known to have recorded Arana. *La Rioja*, September 5, 1894.

118. "Sociedad de Fomento," *Boletín Republicano de la Provincia de Gerona*, April 24, 1897, 3.

119. The capabilities and size of the phonograph at this time also make it very unlikely that the recording was made through bootlegging.

Chapter 3

1. Columbia launched the graphophone and was releasing recording catalogues as early as 1891; however, before 1895 very few individuals, even in America, owned a phonograph for domestic use (Gelatt, *Edison's Fabulous Phonograph*, 44–5, 56, 69).

2. See Michael Kinnear, *The Gramophone Company's First Indian Recordings 1899–1909* (Bombay: Popular Prakashan, 1994), 9.

3. Allan Sutton, *Directory of American Disc Record Brands and Manufacturers, 1891–1943* (Westport, CT, and London: Greenwood Press, 1994), 32–3, 42, 84–5, 178–9.

4. João Silva, *Entertaining Lisbon: Music, Theater, and Modern Life in the Late 19th Century* (New York: Oxford University Press, 2016), 3; Kinnear, *The Gramophone Company's*, 9.

5. For a summary, see Leech-Wilkinson, *The Changing Sound of Music*, chapter 2.

6. Cilindrique, "Fonografía," *El cardo*, November 8, 1900, 17; "Nuestra revista," *El cardo*, November 30, 1900, 20; Cilindrique, "Fonografía madrileña," *El cardo*, April 15, 1901, 14.

7. "Impresiones," *Boletín fonográfico*, 4 (1900), 52–3.

8. Atanasio Palacio Valdés, from La fonográfica madrileña, patented on March 12, 1902 a procedure to impress two or four cylinders at the same time (patent no. 29454). A man called Marcos Guerrero Oliver filed on September 14, 1900 a patent for a cone-shaped cylinder which could be used as a matrix and copied multiple times (patent no. 26568); however, I have not been able to find out who Marcos Guerrero Oliver was and whether he was connected to any *gabinete*.

9. Data about cylinder sales are self-reported, and come solely from Hugens y Acosta, allowing no comparisons across *gabinetes*. Hugens y Acosta announced in November 1900 that they were planning to produce 8,000 cylinders in the coming months to meet demand ("Fonogramas recibidos," *El cardo*, November 30, 1900, 1), and stated in January 1901 that after two years of operation they had sold 30,000 cylinders

("Industria fonográfica," *El cardo*, January 22, 1901, 14–15). The surviving cylinders (mostly at the Biblioteca Nacional de España) offer a similarly incomplete picture: given that wax cylinders were relatively fragile and soon became obsolete, only a very small fraction has made it to our days. The Biblioteca Nacional de España holds some 180 cylinders from Madrid *gabinetes* (with Hugens y Acosta, Viuda de Aramburo, and Álvaro Ureña being the most numerous), followed by 42 from Barcelona and 27 from Valencia. Most of these can be listened to through Biblioteca Digital Hispanica: http://bdh.bne.es/bnesearch/Search.do?destacadas1=Cilindros+d e+cera&home=true&languageView=en. This would, again, suggest that the Madrid *gabinetes* were the highest-producing ones, but we should also consider the possibility that the collection at the Biblioteca Nacional de España does not offer an accurate sample (due to, e.g., the private collectors who initially bought the cylinders and donated them to the Biblioteca having a preference for certain *gabinetes* over others).

10. Suisman, *Selling Sounds. The Commercial Revolution in American Music*, 15.

11. For example, López Piñero's edited history focuses on scientific developments in universities and by independently wealthy amateurs who dedicated their time to science; there is much less of a focus on independent professionals in the applied sciences (José María López Piñero, ed., *La ciencia en la España del siglo XIX* [Madrid: Marcial Pons, 1992]). The only chapter that deals with applied science practitioners is Francisco Javier Puerto Sarmiento, "Ciencia y farmacia en la España decimonónica," 153–92. Pedro Ruiz-Castell talks about manufacturers and resellers of scientific equipment as an important pillar of scientific development in Spain throughout the nineteenth century in Ruiz-Castell, "Scientific Instruments for Education in Early Twentieth-Century Spain," 519.

12. The full name of the *gabinete* was Sociedad Fonográfica Española Hugens y Acosta. It was commonly abbreviated to Hugens y Acosta, but some recordings are catalogued as Sociedad Fonográfica Española, for example at the Biblioteca Digital Hispánica.

13. Mathieu-Villon, *El fonógrafo y sus aplicaciones*.

14. *El diario de Córdoba*, March 21, 1894, 2; *Heraldo de Baleares*, October 19, 1895, 2; *El Globo*, December 16, 1895; *Crónica meridional*, January 31, 1896; *El Guadalete*, March 7, 1896; *El bien público*, October 7, 1896; "Crónica local," *El isleño*, October 16, 1896, 2.

15. "Madrid," *El globo*, March 15, 1897, 2.

16. *El Correo militar,* December 9, 1896, 3.

17. Pedro José Chacón Delgado, *Historia y nación. Costa y el regeneracionismo en el fin de siglo* (Santander: Ediciones de la Universidad de Cantabria, 2013), 51.

18. "Madrid Artístico. Sociedad Fonográfica Española—Hugens y Acosta," *El Liberal*, October 3, 1899, 2.

19. For example, in Lucas Mallada, *La futura revolución española y otros escritos regeneracionistas*, ed. Francisco J. Ayala-Carcedo and Steven L. Driever (Madrid: Biblioteca Nueva, 1998), 139 and 226.

20. *El correo militar*, November 29, 1883, 3; *El correo militar*, March 26, 1890, 3; *El reservista*, January 19, 1895, 3.

21. *ABC*, August 19, 1911, 12.

22. *Diario oficial de avisos de Madrid*, December 6, 1895, 2; *El bien público*, December 7, 1895.
23. "Notas de electricidad," *Madrid científico*, 146 (1897): 10.
24. Erik Swyngedouw, *Liquid Power. Contested Hydro-Modernities in Twentieth-Century Spain* (Cambridge, MA: MIT Press, 2015), 43.
25. *El Ateneo*, July 10, 1898, 7; "El mejor fotógrafo," *El Ateneo*, August 10, 1896, 8.
26. *La música ilustrada hispano-americana*, April 25, 1899, 10.
27. "Nuevo establecimiento," *El correo militar*, July 19, 1897, 2.
28. *La correspondencia militar*, October 12, 1900, 2.
29. "Enrique Sepúlveda," *La época*, December 29, 1898, 1.
30. *La correspondencia de España*, December 19, 1897; "Círculo de la Unión Mercantil," *La época*, December 19, 1897, 2.
31. Naranjo Santana, *Cultura, ciencia y sociabilidad en Las Palmas de Gran Canaria*, 142.
32. Zozaya, *Identidades en juego*, 136.
33. Salavert and Suárez Cortina, "Introducción," 11.
34. Lisa Gitelman, "Reading Music, Reading Records, Reading Race: Musical Copyright and the U.S. Copyright Act of 1909," *The Musical Quarterly*, 81:2 (1997), 265–90 (p. 267); Velásquez Ospina, Juan Fernando, *(Re)sounding Cities*, 259–60.
35. *El correo militar*, July 19, 1897.
36. Álvaro Ureña, *Álvaro Ureña* (advertisement), *La época*, December 15, 1902, 3.
37. "Automóviles en Madrid," *La época*, September 21, 1899, 3.
38. "Exposición madrileña de pequeñas industrias," *El Siglo futuro*, June 10, 1901, 1.
39. *La época*, May 15, 1902, 2.
40. M. [full name unknown], "Crónicas madrileñas," *La época*, December 22, 1904, 1.
41. *La correspondencia de España*, January 2, 1900, 3.
42. Armando Hugens, *Sociedad Fonográfica Española* (advertisement), *Guía de la Coronación hecha expresamente para los forasteros que visiten Madrid en las fiestas que se celebrarán durante el mes de Mayo de 1902 con motivo de la coronación de S.M. el Rey D. Alfonso XIII*, ed. Gerardo Pardo (Madrid: publishing house unknown, 1902), 96–7.
43. [First name unknown] Garci-Fernández, "Política europea," *La región extremeña*, February 8, 1900, 1; [First name unknown] Garci-Fernández, "Política europea," *La región extremeña*, March 1, 1900, 1. For the effects of the early twentieth-century protectionist tax reform on science, see Sánchez Ron, "Investigación científica," 51.
44. *El cardo*, March 8, 1901, 15; Cilindrique, "Cosas fonográficas," *El cardo*, June 8, 1901, 14.
45. See for example El graphos, *El graphos* (advertisement), *La region extremeña*, March 23, 1900, 4; [First name unknown] Garci-Fernández, "Política europea," *Crónica meridional*, January 26, 1900, 1.
46. Viuda de Aramburo, *Viuda de Aramburo* (advertisement), *ABC*, May 22, 1897, 3.
47. The most prominent cases include Fono-Reyna Sociedad Fonográfica Española (10 unnamed cylinders, out of a total of 44 survived) and La fonográfica madrileña (8 unnamed cylinders, out of a total of 27 survived). Out of these 18 unnamed cylinders, 11 are choral or instrumental pieces.
48. Feaster, "Framing the Mechanical Voice," 91.

49. Interestingly, it appears in some homemade recordings too, as will be discussed in Chapter 7 concerning the private recordings of Ruperto Regordosa.

50. Feaster, "'The Following Record'," 305.

51. Feaster, "'The Following Record'," 311–13.

52. The role of sound engineer or producer as such did not formally exist at the time. On sound engineers and producers in early phonography see Blake, "Recording Practices and the Role of the Producer," 36–53; Chanan, *Repeated Takes*, 57.

53. Susan Schmidt-Horning, "Engineering the Performance: Recording Engineers, Tacit Knowledge, and the Art of Controlling Sound," *Social Studies of Science*, 34 (2004), 703–31 (p. 704 and 711).

54. *La Música ilustrada hispano-americana*, April 25, 1899.

55. La fonográfica madrileña, *La fonográfica madrileña* (advertisement), *ABC*, April 2, 1903, 4; La fonográfica madrileña, *La fonográfica madrileña* (advertisement), *El heraldo de Madrid*, May 6, 1902, 4; La fonográfica madrileña, *La fonográfica madrileña* (advertisement), *Nuevo mundo*, March 18, 1903, 2.

56. Marqués de Alta-Villa, "Fonografía. Cuestión palpitante," *El cardo*, April 8, 1901, 14–15.

57. *La época*, February 15, 1900, 2; Álvaro Ureña, "Comunicado," *La correspondencia militar*, February 16, 1900, 2.

58. Ureña, "Comunicado," 2.

59. Cilindrique, "Para todos, sabios e ignorantes," *El cardo*, December 15, 1900, 14.

60. Cilindrique, "Para todos, sabios e ignorantes," 14. At this stage, foreign companies whose cylinders were sold in Spain include Edison, Gramophone, and Lioret.

61. Cilindrique, "Cosas de fonografía," *El cardo*, March 30, 1901, 15–16; Cilindrique, "El arte y el gramófono," *El cardo*, December 22, 1900, 14. The latter article identifies Corrons's in Barcelona and Hugens y Acosta's in Madrid as the best produced cylinders in Spain.

62. "Industria fonográfica," 14. See also Cilindrique, "De fonografía," *El cardo*, September 8, 1901, 14.

63. Cilindrique, "Asuntos fonográficos," *El cardo*, October 8, 1901, 14.

64. Álvaro Ureña, [untitled], *El cardo*, March 8, 1901, 15.

65. Marqués de Alta-Villa, "Fonografía. Cuestión palpitante," 14.

66. Marqués de Alta-Villa, "Fonografía. Cuestión palpitante," 15. See also Cilindrique, "Cosas fonográficas," *El cardo*, June 22, 1901, 14.

67. "Gabinetes fonográficos españoles: Sres. Pallás y Comp., de Valencia," *Boletín fonográfico* 18 (1901), 294–7.

68. *El Cardo*, February 15, 1901, 14; *El cardo*, February 22, 1901, 15.

69. Cilindrique, "Fonografía madrileña," *El cardo*, April 15, 1901, 14.

70. Cilindrique, "Cancanes fonográficos," *El cardo*, November 15, 1900, 22.

71. These calculations take into account the holdings at the Biblioteca Nacional de España, Biblioteca de Catalunya, Museu de la Música, Eresbil, and Centro de Documentación Musical de Andalucía, and concern commercial cylinders only: the collection of Ruperto Regordosa held at the Biblioteca Nacional de Catalunya has not been taken into account.

72. *El cardo*, February 15, 1901, 14. See also "Sobre la originalidad de los cilindros," *El cardo*, March 30, 1901, 14.

73. "El fonógrafo y el gramófono," *El cardo*, February 8, 1901, 14.

74. Cilindrique, "De fonografía," 14.

75. La fonográfica madrileña, *La fonográfica madrileña* (advertisement), *ABC*, April 2, 1903, 4.

76. Between late 1903 and mid-1904, Ureña set up a rather ambitious newspaper advertising campaign in a number of provincial Spanish newspapers, advertising his new identity as a Gramophone distributor. See, for example, Álvaro Ureña, *Álvaro Ureña* (advertisement), *Diario de Córdoba*, May 7, 1904, 4; Álvaro Ureña, *Álvaro Ureña* (advertisement), *El adelanto de Salamanca*, July 31, 1904, 4.

77. Fono-Reyna, *Fono-Reyna* (advertisement), *La correspondencia de España*, August 12, 1903, 4.

78. La fonográfica madrileña, *La fonográfica madrileña* (advertisement), *ABC*, April 2, 1903; La fonográfica madrileña, *La fonográfica madrileña* (advertisement), *ABC*, January 28, 1905.

79. Pekka Gronow and Björn Englund, "Inventing Recorded Music: The Recorded Repertoire in Scandinavia, 1899–1925," *Popular Music*, 26:2 (2007), 281–304 (p. 292).

Chapter 4

1. "Industria fonográfica," *El cardo*, January 22, 1901, 14.

2. Cilindrique, "El arte y el gramófono," 14.

3. Corrons is named in "Gabinete fonográfico," *Boletín fonográfico*, 35–6 (1901), 133. Mentions to Barcelona include: "De re fonográfica," *Boletín fonográfico*, 28 (1901), 69–70; "De re fonográfica," *Boletín fonográfico*, 35–6 (1901), 185–6.

4. Grandes Almacenes El Siglo, *Grandes Almacenes El Siglo* (advertisement), *La Vanguardia*, February 16, 1899, 7.

5. "Hojas sueltas," *Los deportes*, April 15, 1899, 142; see also *La música ilustrada hispano-americana*, April 25, 1899, 10.

6. "Teatros," *La publicidad*, April 21, 1899, 4.

7. *La publicidad*, April 23, 1899, 3.

8. "El fonógrafo ambulante," *La dinastía*, May 15, 1899, 1.

9. "Crónica," *La hormiga de oro*, April 22, 1899, 15.

10. *Industria e invenciones*, April 22, 1899, 155–6.

11. These include no. 25373 from January 1900 for modifications to the Mignon phonograph; no. 39964 in January 1907 for a diaphragm; no. 51102 in August 1911 for improvements in diaphragms.

12. The cylinder is held at the Biblioteca Nacional de España; it features tenor Ángel Constantí singing "Viva il vino" from *Cavalleria rusticana*.

13. In October 1899, Rosillo filed an application with the municipal council to paint the façade of the shop; see Fo-783-AH at the Arxiu Municipal de Barcelona.

14. "Notas locales," *La vanguardia*, July 21, 1899, 2.

15. *Heraldo de Alcoy*, April 30, 1901, 3.

16. [First name unknown] Roselló, *Óptico Roselló* (advertisement), *La Vanguardia*, May 7, 1899, 1; [First name unknown] Roselló, *Óptico Roselló* (advertisement), *Las Noticias*, May 7, 1899, 2. The establishment had been open until that name since at least 1897, as it appeared in the *Anuario Riera* for that year.

17. [First name unknown] Roselló, *Óptico Roselló* (advertisement), *La Vanguardia*, April 18, 1898, 3.

18. Juan Bautista Estradé Simó and Ricardo Ribas Anguera, *Ribas y Estradé* (advertisement), *La Vanguardia*, October 31, 1899, 3.

19. It is likely that Ribas y Estradé or other Barcelona *gabinetes* benefitted from or at least knew about the research on beeswax that was being carried out around this time at the School of Pharmacy of the University of Barcelona; see Casimiro Brugués y Escuder, *La cera de abejas: procedimientos para ensayar la cera de abejas y su aplicación al examen de diferentes muestras de ceras españolas procedentes de apicultores, de farmacias, de rito, de droguerías y del comercio de cerería* (Barcelona: Imprenta de Pedro Ortega, 1900).

20. *La Corona*, April 8, 1863, 2.

21. Gómez-Montejano, *El fonógrafo en España*, 101.

22. José Corrons, *Óptica Corrons* (advertisement), *La Vanguardia*, December 13, 1900, 6.

23. Quoted in Javier Barreiro and Gabriel Marro, *Primeras grabaciones fonográficas en Aragón*, 10.

24. Mariano Gómez-Montejano, "El fonógrafo en España (tercera parte) ," *Girant a 78 rpm*, 8 (2005), 6–8.

25. José Corrons, *Canals & Corrons* (advertisement), *Boletín fonográfico*, 35–6 (1901), 194.

26. Grammophon, *Grammophon. Verzeichnis französischer italienischer spanischer portugiesischer Platten*, no date (estimated ca. 1900), reference number Cat 1/2.

27. Manuel Moreno Casas, *Centro fonográfico comercial* (advertisement), *La Vanguardia*, November 15, 1900, 8.

28. O. [first name unknown] Pernat, *O. Pernot* [sic] (advertisement), *La Vanguardia*, January 13, 1901, 3. Whereas in the first advertisement published in the press the name of the *gabinete* is given as O. Pernot, in subsequent advertisements it appears as O. Pernat, suggesting that the first one was a mistake.

29. These include three recordings at Biblioteca Digital Hispánica and a further six at the Museu de la Mùsica in Barcelona. In the Museu's catalogue, the cylinders are dated as 1890—but this seems unlikely, given that no company in Spain was manufacturing cylinders for commercial use at that time.

30. Oliver Hochadel and Laura Valls, "Civic Nature. The Transformation of the Parc de la Ciutadella into a Space for Popular Science," *Barcelona: An Urban History of Science and Modernity, 1888–1929*, ed. Oliver Hochadel and Agustí Nieto-Catalán (Abingdon: Routledge, 2016), 25–45.

31. Joan Ramon Resina, *La vocació de modernitat de Barcelona: auge i declivi d'una imatge urbana* (Barcelona: Galàxia Gutenberg, 2008), 55.

32. Resina, *La vocació de modernitat*, 59.

33. Resina, *La vocació de modernitat*, 15.

34. Ferran Sagarra, "Barcelona dins del projecte industrialista català," *La formació de l'Eixample de Barcelona: aproximacions a un fenomen urbà*, ed. Santi Barjau et al. (Barcelona: Olimpíada Cultural, 1990), 13–25 (p. 17).

35. Miquel Corominas, "Les societats de l'eixample," *La formació de l'Eixample de Barcelona: aproximacions a un fenomen urbà*, ed. Santi Barjau et al. (Barcelona: Olimpíada Cultural, 1990), 43–59 (p. 45).

36. Albert Garcia Espuche, *El Quadrat d'Or: centro de la Barcelona modernista: la formación de un espacio urbano privilegiado* (Barcelona: Ajuntament de Barcelona, Lunwerg Editores, 2002), 11.

37. Garcia Espuche, *El Quadrat d'Or*, 16 and 19.

38. Garcia Espuche, *El Quadrat d'Or*, 21.

39. Garcia Espuche, *El Quadrat d'Or*, 45.

40. Cèlia Cañellas and Rosa Torán, "Heterogeneïtat urbana, desplaçaments geogràfics i canvis polítics," *La formació de l'Eixample de Barcelona: aproximacions a un fenomen urbà*, ed. Santi Barjau et al. (Barcelona: Olimpíada Cultural, 1990), 189–202 (p. 192).

41. Albert Garcia Espuche, "El centre residencial burgès (1860–1914)," *La formació de l'Eixample de Barcelona: aproximacions a un fenomen urbà*, ed. Santi Barjau et al. (Barcelona: Olimpíada Cultural, 1990), 203–21.

42. Garcia Espuche, *El Quadrat d'Or*, 69.

43. Garcia Espuche, *El Quadrat d'Or*, 85; Cañellas and Torán, "Heterogeneïtat urbana," 197.

44. Garcia Espuche, *El Quadrat d'Or*, 101.

45. Garcia Espuche, *El Quadrat d'Or*, 77.

46. Garcia Espuche, *El Quadrat d'Or*, 107.

47. Jordi Casassas, *La voluntat i la quimera. El noucentisme català entre la Renaixença i el marxisme* (Barcelona: Pòrtic, 2017).

48. José Luis Oyón, *La quiebra de la ciudad popular: espacio urbano, inmigración y anarquismo en la Barcelona de entreguerras, 1914–1936* (Barcelona: Ediciones del Serbal, 2008), 9–25. See also Angel Smith, "From Subordination to Contestation: The Rise of Labour in Barcelona, 1898–1918," *Red Barcelona. Social Protest and Labour Mobilization in the 20th century*, ed. Angel Smith (London: Routledge, 2002), 17–43.

49. Pere Gabriel, "Red Barcelona in the Europe of War and Revolution, 1914–1930," *Red Barcelona. Social Protest and Labour Mobilization in the 20th Century*, ed. Angel Smith (London: Routledge, 2002), 44–65 (p. 48).

50. *La Vanguardia*, December 27, 1892, 3.

51. Arxiu Municipal de Barcelona, file no. AMCB1-001, 1892.

52. See also Antoni Torrent i Marqués, "Efectes secundaris dels discs a 78 rpm," *Girant a 78 rpm*, 2:4 (2004), 5–12. Torrent i Marqués makes a similar point that the most "up-to-date" activity around 1900 is found in the Casc Antic, Barri Gotic, and Rambles, and this progressively grows toward the North. However, it is not clear whether he is referring to the *gabinetes fonográficos* here as he seemed to have only limited

knowledge of them; it is perhaps more likely that he refers to the first record shops importing from other countries.

53. Building plans, file no. 21596, 1878, Arxiu Municipal de Barcelona.

54. Resina, *La vocació de modernitat*, 82.

55. Garcia Espuche, *El Quadrat d'Or*, 7.

56. Raffaella Perrone, "Espacio teatral y escenario urbano. Barcelona entre 1840 y 1923" (Ph.D. diss., Universitat Politècnica de Catalunya, 2011), 28.

57. Perrone, *Espacio teatral y escenario urbano*, 90–1.

58. Juan Rosillo, *Sociedad artístico-fonográfica* (advertisement), *La Publicidad*, June 6, 1899, 3. No indication is given as to which excerpts of the opera were recorded and who recorded them.

59. [First name unknown] Roselló, *Óptico Roselló* (advertisement), *La Vanguardia*, May 7, 1899, 3.

60. The same is true for the two surviving cylinders from Grandes Almacenes El Siglo, although here the sample is too small to extrapolate.

61. Hochadel and ANieto-Catalán, "Introduction," 13; Girón Sierra Jorge Molero-Mesa, "The Rose of Fire," 115–35.

62. Among these new formats, Hochadel and Nieto-Catalán list: "public museums, urban parks, international exhibitions, electric lighting, private clinics, amusement parks, newspapers, evening schools, new places of sociability such as athenaeum, cinema, radio, anarchist and spiritist circles" —but, interestingly, not recording technologies in any form; tellingly, they do not list the phonograph or other sound technologies, suggesting that their role was not as significant as the other previously mentioned practices. Hochadel and Nieto-Catalán, "Introduction," 5. See also: Jaume Sastre-Juan and Jaume Valentines-Álvarez, "Technological Fun. The Politics and Geographies of Amusement Parks," *Barcelona: An Urban History of Science and Modernity, 1888–1929*, ed. Oliver Hochadel and Agustí Nieto-Catalán (Abingdon: Routledge, 2016), 115–35.

63. Hochadel and Nieto-Catalán, "Introduction," 2.

64. Casassas, *La voluntad y la quimera*, 59–60; Angel Smith, "Barcelona through the European Mirror: From Red and Black to Claret and Blue," *Red Barcelona. Social Protest and Labour Mobilization in the 20th Century*, ed. Angel Smith (London: Routledge, 2002), 1–16 (p. 2).

65. Fèlix Villagrasa i Hernàndez, *Mancomunitat i ciència: la modernització de la cultura catalana* (Catarroja/Barcelona/Palma: Afers, 2015); Santiago Riera i Tuèbols, *Història de la ciència a la Catalunya moderna* (Lleida i Vic: Pagès/Eumo, 2003).

66. Antoni Roca Rosell, "El discurso civil en torno a la ciencia y la técnica," *El regeneracionismo en España. Política, educación, ciencia y sociedad*, ed. Vicente Salavert and Manuel Suárez Cortina (Valencia: Universitat de Valencia, 2007), 241–59 (p. 247).

67. A is the case with Spain, the historiography of science in nineteenth-century Catalonia has tended to focus on academic science and pure research. See for example Josep M. Camarasa y Antoni Roca Rosell, eds., *Ciència i tècnica als Països Catalans: una aproximació biogràfica* (Barcelona: Fundació Catalana per a la Recerca, 1995); Hochadel and Nieto-Catalán, "Introduction," 14.

68. *Exposición universal de Barcelona*, p. 34.

69. For example, Isidro Torres y Oriol, *Barcelona histórica antigua y moderna: guía general descriptiva e ilustrada* (Diputación y Ayuntamiento, 1905).

70. *Recopilacion de estudios e investigaciones efectuada por la Comisión Obrera Catalana en la Exposición de Chicago* (Barcelona: unknown publisher, 1893).

71. Perramon i Basqué, "Presentació de l'Aspe," 3–4; Torrent i Marqués, "Girant al voltant dels 'Berliner's,'" 3–7.

72. In fact, the piano roll industry in Barcelona is still to be examined from the point of view of the cultural discourses it disseminated, as well as the listening practices it introduced; such a study will likely reveal interesting concomitances with contexts in which piano rolls "mediated between the 'familiar' timbre of the piano and the disembodied sounds of the phonograph" and inserted themselves within discourses of democratization and civilization; see Ospina Romero, Sergio, "Ghosts in the Machine and Other Tales," 5–7.

73. Alan Kelly, *The Gramophone Company Limited, The Spanish Catalogue: Including Portuguese Recordings* (London: self-published, 2006). Please note that no page numbers are given as the book does not have them. See also Antoni Torrent i Marqués, "Una recerca interessant (1° part)," *Girant a 78 rpm*, 8 (2005), 9–11.

74. Torrent i Marqués, "Una recerca interessant (1° part)," 10.

75. Antoni Torrent i Marqués, "Visió europea del naixement de l'enregistrament sonor," *Girant a 78 rpm*, 1:1 (2002), 5–9 (p. 6). Antoni Torrent i Marqués, "Girant al voltant dels 'Berliner's,'" 5.

76. Torrent i Marqués, "Visió europea," 8.

77. Rainer E. Lotz, "On the History of Lindström AG," *The Lindström Project*, vol. 1, ed. Pekka Gronow and Christiane Hofer (Vienna: Gesellschaft für Historische Tonträger, 2009), 11–22.

78. Richard K. Spottswood, *Ethnic Music on Records: A Discography of Ethnic Recordings Produced in the United States, 1893 to 1942* (Urbana and Chicago: University of Illinois Press, 1990), 1:xv.

79. Eva Moreda Rodríguez, "Recording *zarzuela grande* in Spain in the Early Days of the Phonograph and Gramophone," *Music, Nation and Region in the Iberian Peninsula*, ed. Samuel Llano, Matthew Machin-Autenrieth, and Salwa Castelo-Branco (Champaign: University of Illinois Press, forthcoming in 2021); John R. Bolig, *The Victor Discography. Green, Blue, and Purple Labels (1910–1926)* (Denver: Mainspring Press, 2006), vii.

80. Seven records were released, containing nine pieces, as follows: *Cançó de Nadal* (Pérez) / *Lo cant dels aucells* (Millet); *L'hereu Riera / Lo Rossinyol*; *Sota de l'olm/L'emigrant*; *Les flors de Maig* (first and second part); *Aucellada* (first and second part); *Els xiquets de Valls* (first and second part); *Entre flors* (Nicolau) / *O magnum misterium* (Victoria). The minutes of the Orfeó Catalá suggest that only a small number of singers were involved in the recording (minutes of the Orfeó Catalá, September 16, 1916, Biblioteca de Catalunya).

81. Gramophone / His Master's Voice, *Orfeó Catalá* (catalogue), 1916 (Biblioteca Nacional de Catalunya, Cat 8/11).

82. John R. Bolig, *The Victor Red Seal Discography vol. II. Double-Sided Series to 1930* (Denver: Mainspring Press, 2004).

83. Antoni Torrent i Marqués, "Una recerca interessant (1º part)," *Girant a 78 rpm*, 8 (2005), 9–11.

84. Antoni Torrent i Marqués, "Els primers enregistraments a casa nostra," 9.

85. Antoni Torrent i Marqués, "Sardanes a 78 rpm. Enregistraments durant els anys 1906–1907," *Girant a 78 rpm*, 15 (2009), 13–17 (p. 13).

86. Torrent i Marqués, "Sardanes a 78 rpm," 14.

87. Torrent i Marqués, "Sardanes a 78 rpm," 16.

88. Torrent i Marqués, "Els primers enregistraments a casa nostra," 9.

89. Torrent i Marqués, "Efectes secundaris dels discs a 78 rpm," 7.

90. José María Folch y Torres, *Barcelona: itinerarios para visitar la ciudad, Exposición Internacional de Industrias Eléctricas y Nacional de Productos: guía práctica para el turismo* (Barcelona: Thomas, 1914), xiii.

91. Velasquez Ospina, Juan Fernando, *(Re)sounding Cities*, 236–8.

92. Museu de la Mùsica, Barcelona, ítem no. MDMB 1211.

93. In *Anuario del comercio*, 1905, Vicente still appears as the owner of a business selling taps and lamps, but in July 1907 it was already supplying phonographs to Valencia; see Eusebio Vicente, *Compañía de máquinas parlantes* (advertisement), *El Pueblo*, July 1, 1907, no page number.

94. Among the former: MDMB 1224 at the Museu de la Mùsica, Barcelona; among the latter: MDMB 799 at the Museu de la Mùsica, Barcelona; MFM S-26006, Museu Federic Marès, Barcelona.

95. Museu de la Mùsica, Barcelona, catalogue no. MDMB 1023.

96. Museu de la Mùsica, Barcelona, catalogue no. MDMB 1140.

Chapter 5

1. This figure does not include some 280 cylinders, most of which held at Eresbil, which have arrived at us in boxes with no indication of the *gabinete* they were recorded in and which are to be too deteriorated to be played back or digitized (in which case the name of the *gabinete* would likely be given in the opening announcement). Nevertheless, many of these cylinders are by performers who we know recorded exclusively for the Valencia *gabinetes* (e.g., Lamberto Alonso, María Vendrell, and Sexteto Goñi for Hijos de Blas Cuesta; Gabriel Hernández and José Bellver for Puerto y Novella), so the total list of surviving cylinders from cities other than Madrid and Barcelona is likely to be higher than indicated.

2. About twenty cylinders recorded by *gabinetes* outside the main three cities have survived. This does not include cylinders labeled by the *gabinete* Viuda de Ablanedo in Bilbao: as will be discussed subsequently, it is likely that most or all of those cylinders were made by Viuda de Aramburo in Madrid, with Ablanedo simply acting as a reseller.

3. This is the case, for example, with Prudencio Santos Benito in Salamanca: there are numerous advertisements in the local press of his department store in the center of the city, but none of them mentions phonographs or cylinders; we know that he might have occasionally impressed some because one—Antonio Vargas singing *La Marsellaise* in Spanish—has survived in the private collection of Mariano Gómez-Montejano; it is digitized at Biblioteca de Catalunya, *Cilindres de cera de la col·lecció Mariano Gómez Montejano*. Similarly, *El cardo* included Málaga among the cities which had *gabinetes fonográficos* in the timespan from late 1900 to early 1901; however, no cylinders or written sources have survived indicating what their names were; see "Concurso fonográfico," *El cardo*, December 15, 1900, 14; "Industria fonográfica," 14.

4. Barreiro and Marro, *Primeras grabaciones fonográficas en Aragón*, 25.

5. Carlos Martín Ballester, "Frequently Asked Questions," http://www.carlosmb.com/pruebas/e_preguntas.php, accessed December 2018.

6. I am grateful to Jaione Landaberea, sound curator at Eresbil, for her assistance with this matter.

7. I.C. [full name unknown], "Desde Béjar," *El adelanto*, January 8, 1906, 1.

8. Patent number 25305, filed on December 30, 1899.

9. "Noticias," *La voz de la provincia*, April 8, 1897, 2.

10. Enrique García's cylinders were all recorded by a señor Arriaga of whom little is known (all cylinders held at Eresbil, Fondo Familia Ybarra, not catalogued). La Oriental's surviving cylinders are held at the Biblioteca Nacional de España and include one *jota* by Balbino Orensanz (CL/299), another by an unknown performer (CL/296), and an ocarina duet by Orensanz and a señor Lahuerta (CL/297). Balbino Orensanz was one of the main *jota* performers of his time, and Barreiro and Marro claim that La Oriental might have also recorded another prominent *jotero*, Santiago Lapuente (Barreiro and Marro, *Primeras grabaciones fonográficas en Aragón. 1898–1903*, 23).

11. "Por lo flamenco," *Boletín fonográfico*, 10 (1900), 156–8; un patriota [pseudonym], "Fonograma," *El cardo*, February 28, 1901, 15.

12. Bolig, *The Victor Discography. Green, Blue, and Purple Labels*, vii; Spottswood, *Ethnic Music on Records*, 1:xv.

13. Denning, *Noise Uprising: The Audiopolitics of a World Musical Revolution*, 1.

14. "Gabinetes fonográficos españoles. II. Sres. Hijos de Blas Cuesta, de Valencia," *Boletín fonográfico*, 9 (1900), 140–2.

15. Manuel Torres Orive, "El fonógrafo," *Las provincias*, June 30, 1899, 2.

16. "Manuel Pallás," *Boletín fonográfico*, 2 (1900), 22.

17. "Gabinetes fonográficos españoles. Sres. Pallás y Comp., de Valencia," *Boletín fonográfico*, 18 (1900), 294–7 (p. 295).

18. Although the date when Hércules Hermanos started operation is unknown, it was certainly operative by the end of 1899, as were the other three *gabinetes* in the city. See "El fonógrafo en Valencia," *Las provincias*, December 2, 1899, 2.

19. "Gabinetes fonográficos españoles. VI. Sres. Hércules Hermanos de Valencia," *Boletín fonográfico*, 23 (1900), 383–5 (p. 384).

20. [First name unknown] Vinaixa, "Noticias," *El liberal*, October 19, 1899, 2.

21. "Gabinetes fonográficos españoles. Sres. Puerto y Novella, de Valencia," *Boletín fonográfico*, 14 (1900), 218–22 (p. 218).

22. Jacobo Muñoz Duato, "Vicente Gómez Novella, 1871–1956" (Ph.D. diss., University of Valencia, 2015); "Gabinetes fonográficos españoles. Sres. Puerto y Novella, de Valencia," 218.

23. Federico Martínez Roda, *Valencia y las Valencias: su historia contemporánea (1800–1975)* (Valencia: Fundación Universitaria San Pablo CEU, 1998), 78–9, 84, 87.

24. Martínez Roda, *Valencia y las Valencias*, 104.

25. Martínez Roda, *Valencia y las Valencias*, 39–40, 46.

26. Martínez Roda, *Valencia y las Valencias*, 193 and 338.

27. For example, Hércules Hermanos was apparently known for its low prices ("De re fonográfica," *Boletín fonográfico*, 29 (1901), 84.

28. "A los lectores," *Boletín fonográfico*, 1 (1900), 3.

29. These include A.[first name unknown] Marín, "La impression de fonogramas," *Boletín fonográfico*, 2 (1900), 19–22; A. [first name unknown] Taltavull, "Bocinas y tubos," *Boletín fonográfico*, 11 (1900), 168–70; A. T[altavull], "Instrucciones especiales para el manejo del fonógrafo," *Boletín fonográfico*, 22 (1900), 362–3; A. T[altavull], "El cilindro fonográfico," *Boletín fonográfico*, 30 (1901), 101–2; "Impresión de cilindros," *Boletín fonográfico*, 32 (1901), 134–5.

30. "Francisca Segura," *Boletín fonográfico*, 26 (1901), 36.

31. *Guía práctica de Valencia* (Valencia, 1898: Imprenta de José Ortega), 112–13.

32. "El maestro Bellver," *Boletín fonográfico*, 6 (1900), 81.

33. "Gabinetes fonográficos españoles. Sres. Puerto y Novella, de Valencia," 222.

34. J. B. [full name unknown], "Carmelo Bueso Beltrán," *Boletín fonográfico*, 8 (1900), 120.

35. "El maestro Rodríguez," *Boletín fonográfico*, 13 (1900), 204.

36. "El maestro Goñi," *Boletín fonográfico*, 22 (1900), 364. The surviving cylinders are both held at Eresbil and are both uncatalogued and undigitized; these are labeled as Brahms's *Danza húngara* (no indication of number) and *Las escopetas*, which is likely an instrumental arrangement from the *zarzuela* of the same title, composed by Ramón Estellés and Joaquín Valverde and premiered in 1896.

37. "Gabinetes fonográficos españoles. VI. Sres. Hércules Hermanos de Valencia", 384.

38. "Maximilano Thous," *Boletín fonográfico*, 3 (1900), 33.

39. "José Bayarri," *Boletín fonográfico*, 15 (1900), 248.

40. "Gabinetes fonográficos españoles. D. José Navarro, de Madrid," *Boletín fonográfico*, 34 (1901), 167–9 (p. 167).

41. Repertoire is discussed in more detail in Chapter 6.

42. "Gabinetes fonográficos españoles. D. José Navarro, de Madrid," 167.

43. A. T[altavull], "Impresión de cilindros," *Boletín fonográfico*, X (1901), 150.

44. "Diafragma The Keating. Reproductor para gramófonos," *Boletín fonográfico*, 1 (1900), 6–7.

45. Martínez Roda, *Valencia y las Valencias*, 196.

46. José Alcañiz, *José Alcañiz* (Advertisement), *Boletín fonográfico*, 3 (1900), 7.

47. *Gran feria de Valencia. 20 al 31 de junio de 1903 (programa de fiestas)* (festival booklet) (Valencia: no publisher, 1903).

48. "Sres. Fiol y Villar," *Boletín fonográfico*, 6 (1900), 87. See also patent no. 24267, from May 23, 1899.

49. Casa Cabedo, *Graduador de diafragma* (advertisement), *Boletín fonográfico*, 9 (1900), 131; see also patents 24805 from September 21, 1899 and 25210 from December 11, 1899.

50. Casa Cabedo, *Novedades musicales para impresionar cilindros* (Advertisement), *Boletín fonográfico*, 9 (1900), 145.

51. Cilindrique, "Cancanes fonográficos," 22. The author does not indicate which *gabinetes* he was referring to, but he singled Hugens y Acosta and "an optician at Calle del Príncipe" (likely Obdulio B. Villasante) as the only ones who had technical know-how in the early days of the recording industry in Madrid.

52. "Gabinetes fonográficos españoles," 140.

53. "Gabinetes fonográficos españoles. Sres. Puerto y Novella, de Valencia," 219.

54. "Los aranceles y el fonógrafo," *Boletín fonográfico*, 3 (1900), 36.

55. "El mercado fonográfico," *Boletín fonográfico*, 23 (1901), 382.

56. "El mercado fonográfico," *Boletín fonográfico*, 14 (1900), 214.

57. "De re fonográfica," 69–70.

58. "Aquí y fuera de aquí," *El cardo*, November 15, 1900, 20.

59. "Valencia," *Las provincias*, April 1, 1899, 2.

60. "Gabinetes fonográficos españoles II. Señores Hijos de Blas Cuesta, de Valencia," 140.

61. "El fonógrafo en Valencia," 2.

62. Mentions to Puerto y Novella representatives can be found in Reus ("Crónica local y general," *Diario de Reus*, March 27, 1900, 2), Tortosa ("Noticias," *El eco de la fusión*, March 22, 1900, 3), and Córdoba ("Fonógrafos," *Diario de Córdoba*, November 21, 1901, 4), whereas Blas Cuesta is known to have had agents and representatives in Tortosa ("Los maestros," *Diario de Tortosa*, December 20, 1899, 2), Mallorca ("Miscelánea," *Mallorca*, July 25, 1899, 432), Jerez ("Anuncios de interés," *El Guadalete*, March 17, 1900, 3), and Santiago de Compostela ("Santiago," *El eco de Santiago*, October 20, 1900, 2).

63. "Gabinetes fonográficos españoles. Sres. Puerto y Novella, de Valencia," 219.

64. "Gabinetes fonográficos españoles. Sres. Pallás y Comp., de Valencia," 294–7.

65. The *salón de impresionar* at Hércules Hermanos looked similar: see "Gabinete fonográfico de los señores Hércules Hermanos. Salón de impresionar," *Boletín fonográfico*, 23 (1900), 373.

66. "Gabinetes fonográficos españoles. Sres. Puerto y Novella, de Valencia," 219.

67. "Gabinetes fonográficos españoles. Sres. Puerto y Novella, de Valencia," 220. Although no photograph of the studio for wind bands is provided, the fact that they were recorded indoors suggests that the size of the ensembles they recorded would be rather reduced. Other *gabinetes*, such as José Navarro's in Madrid, recorded wind bands outdoors; see "Gabinetes fonográficos españoles. D. José Navarro, de Madrid," 168.

68. "Gabinetes fonográficos españoles. Sres. Puerto y Novella, de Valencia," 219.

69. "Noticias," *Boletín fonográfico*, 19 (1900), 297.

70. Vicente Blasco Ibáñez, "Viñas en el fonógrafo," *Boletín fonográfico*, 17 (1900), 279–82 (p. 280). Puerto y Novella advertised Viñas cylinders (at the rather steep price of 40 *pesetas*) ("Catálogo general Puerto y Novella," *Boletín fonográfico*, 14 (1900), 223–9

(p. 224). None of those has survived, but 68 recordings by Viñas for Gramophone & Typewriter (1903) did.

71. Timothy Day, *A Century of Recorded Music: Listening to Musical History* (New Heaven, CT, and London: Yale University Press, 2002), 47.

72. "El diafragma Pallás," *Boletín fonográfico*, 6 (1900), 90.

73. Hijos de Blas Cuesta, *Hijos de Blas Cuesta* (advertisement), *Boletín fonográfico*, 27 (1901), 42.

74. "Los aranceles y el fonógrafo," 35–6. The anonymous writer claimed that import taxes practically doubled the prices of phonographs, and that the taxes for blank cylinders were even more extortionate: one kilogram of cylinders costed 2.5 *pesetas* from the factory, but incurred import taxes of 8 *pesetas*. A further article claimed that Viuda de Aramburo had paid 200,000 *pesetas* in import taxes the previous year ("Impresiones," *Boletín fonográfico*, 4 (1900), 52–3). Similar views were expressed in "Novedades," *El cardo*, March 8, 1901, 15.

75. "Una real orden," *Boletín fonográfico*, 4 (1900), 51–2.

76. "Noticias," *Boletín fonográfico*, 16 (1900), 265.

77. C.[first name incomplete], "Industrias fonográficas," *Boletín fonográfico*, 33 (1901), 148–9.

78. Vinaixa, "Noticias," 2.

79. "Gabinetes fonográficos españoles. Sres. Puerto y Novella, de Valencia," 220.

80. "De re fonográfica," *Boletín fonográfico*, 35–6 (1901), 185–6 (p. 185).

81. "Noticias," *Boletín fonográfico*, 29 (1901), 85.

82. "Fonógrafos," 4.

83. Cilindrique, "Sección fonográfica," *El cardo*, October 8, 1901, 14.

84. *Programa de festejos. Gran feria en la ciudad de Valencia* (Valencia: no publisher, 1903); M. [first name unknown] Gómez, *Guía comercial de Valencia, Alicante y Castellón de la Plana* (Valencia: no publisher, 1904), 270 and 277.

85. T. [full name unknown], "El gramófono," *Las provincias*, March 22, 1904, 2.

86. "El gramófono," 2.

87. "El gramófono," 2.

88. "De re fonográfica," *Boletín fonográfico*, 25 (1901), 14.

89. Gramophone, *El Gramophone* (advertisement), *La atalaya*, August 5, 1903, 4.

90. Numerous surviving Gramophone discs at the Biblioteca Nacional de España also have Prudencio Santos Benito's label on them (e.g., DS/10908/10, DS/10908/2).

91. "Sueltos," *La opinión*, August 30, 1903, 2.

92. Gramophone, *Gramophone Noticias* (advertisement), *El porvenir segoviano*, February 24, 1904, 4.

Chapter 6

1. Sterne, *The Audible Past*, 50.

2. Theodor W. Adorno, "The Curves of the Needle," trans. Thomas Y. Levin, *October*, 55 (1990), 48–55 (p. 54).

3. Picker, "The Victorian Aura of the Recorded Voice," 770; Picker, *Victorian Soundscapes*, 111, 123, 129, 138.

4. Marín, "La impresión de fonogramas," 19–22. A further article at *Boletín fonográfico*, under the pretense of giving advice to record cylinders generally, again focused exclusively on vocal ones: anonymous, "Impresión de cilindros," 134–5.

5. Marín, "La impresión de fonogramas," 20.

6. Gitelman, *Scripts, Grooves, and Writing Machines*, 64; see also David Suisman, "Sound, Knowledge, and the 'Immanence of Human Failure': Rethinking Musical Mechanization through the Phonograph, the Player-Piano, and the Piano," *Social Text*, 28 (2010), 12–34 (p. 23).

7. Thomas A. Edison, *Description of the Phonograph and Phonograph-Graphophone by Their Respective Inventors. Testimonials as to Their Practical Use* (Edison: New York, 1888).

8. Solorno, "El fonógrafo," 53–5; Sorarrain, "La ciencia actual," 3; Heredia y Larrea, *El testamento fonográfico*; Mathieu-Villon, *El fonógrafo y sus aplicaciones*; Blanca Valmont, "Crónica," *La última moda*, 269 (1893), 1–2.

9. An illustrative example of how different uses of the phonograph proliferated in different countries, and the role that public institutions had in steering those one way or another, comes from the postal industry in Mexico, where recorded messages became widespread in the 1890s; see Jaddiel Díaz Frene, "A las palabras ya no se las lleva el viento: apuntes para una historia cultural del fonógrafo en México (1876-1924)," *Historia mexicana*, 66 (2016), 257–98. Also Vest, *Vox Machinae: Phonographs and the Birth of Sonic Modernity*, 81 (on the United States).

10. Gramophone, *Libro de oro de la Compañía Francesa del Gramófono* (Barcelona: Gramophone, 1904).

11. Gelatt, *Edison's Fabulous Phonograph 1877–1977*, 55 and 102; Michael Seil, "Opera Singers on the Lindström Labels," *The Lindström Project*, vol. 1, ed. Pekka Gronow and Christiane Hofer (Vienna: Gesellschaft für Historische Tonträger, 2009), 41–2 (see p. 41); Alexandra Wilson, "Galli-Curci Comes to Town," *The Arts of the Prima Donna in the Long Nineteenth Century*, ed. Rachel Cowgill and Hilary Poriss (New York: Oxford University Press, 2012), 328–47 (p. 330).

12. Untitled drawing, *El Cardo*, August 8, 1901, 13.

13. Edison Bell, *Up to Date Dealer's Special Parcel* (Catalogue) (unknown location, 1906).

14. Sociedad Fonográfica Española Hugens y Acosta, *Catálogo* (Madrid, 1900).

15. Sociedad Fonográfica Española Hugens y Acosta, *Catálogo*.

16. Martín de Vargas, "Julián Biel," *El heraldo de Madrid*, September 21, 1899, 3.

17. Vargas, "Julián Biel," 3.

18. Six cylinders by Julián Biel for Hugens y Acosta have survived (of arias from *Carmen*, *Pagliacci*, *Marina*, and *L'africaine*); due to deterioration, however, none of them can presently be listened to.

19. Vargas, "Julián Biel," 3.

20. Sociedad Fonográfica Española Hugens y Acosta, *Catálogo*, 33.

21. Armando Gresca, "Crónica teatral," *El arte del teatro*, 28 (1907), 1.

22. "Jesús Valiente," *Boletín fonográfico*, 12 (1900), 189.

23. Fernando Sánchez Rebanal, "La vida escénica en la ciudad de Santander entre 1895 y 1904" (Ph.D. diss., Universidad Nacional de Educación a Distancia, 2014), 268.

24. The singing profession in *zarzuela* has not yet been the object of any dedicated study. The numerous recent research on theatrical life in the provinces indeed provides insights into the working lives of singers. The main source for the preceding discussion of opera and *zarzuela* companies is Emilio Casares, "Compañías," *Diccionario de la zarzuela. España e Hispanoamérica*, ed. Emilio Casares (Madrid: ICCMU, 2008), 1:525–39.

25. *Género ínfimo* and *cuplé* developed in the decade 1900–1910 as *género chico* slowly started to decay. Both were close to variety-type shows, with provocative dancing, nudity, and innuendo taking a prominent role. Whereas *género ínfimo* still resembled *género chico* in that it consisted of several musical numbers and spoken dialogue connected by a plot, however loosely defined, a *cuplé* is an autonomous song not integrated in a play, although it is still connected to theatrical genres by the importance given to text delivery, acting, and stage movement.

26. José Deleito y Piñuela, *Origen y apogeo del "género chico"* (Madrid: Revista de Occidente, 1949), 36; Matilde Muñoz, *Historia de la zarzuela y el género chico* (Madrid: Tesoro, 1946), 268–9.

27. These include a duet from *Los dineros del sacristan*, with a tenor called señor Fidel, for Hugens y Acosta (Biblioteca Nacional de España, CL/28); duet from *El año pasado por agua* with Emilio Cabello for Puerto y Novella (Eresbil Archivo de la Música Vasca, FA60/296); a duet from *El puñao de rosas* with Emilio Cabello (Vicente Miralles Segarra collection). A further cylinder, a tercet from *El año pasado por agua*, is held at Eresbil but is too fragile to be digitized.

28. Emilio Casares, "Leocadia Alba," *Diccionario de la zarzuela. España e Hispanoamética*, ed. Emilio Casares (Madrid: ICCMU, 2008), 1:27; Deleito y Piñuela, *Origen y apogeo del "género chico"*, 36; Muñoz, *Historia de la zarzuela y el género chico*, 268–9.

29. Villasante, Cilindros Fonográficos J. Oliva and Viuda de Aramburo were a mere few yards from the Teatro de la Comedia), La fonográfica madrileña a turn away from Teatro de la Zarzuela, and Hugens y Acosta right next to the exit of Teatro Apolo in Calle del Barquillo.

30. A further cylinder (the song *Córdoba: canción española*, recorded for Viuda de Aramburo) is catalogued at the Biblioteca Nacional de España as performed by Sebastián Bezares (CL/167). There are no records, however, of any singer of that name, suggesting that this might be a mistake and the cylinder might indeed be by Rafael Bezares.

31. Christián de Neuvillette, "Crónicas de sociedad," *La correspondencia de España*, August 26, 1901, 2.

32. "Nuestros artistas," *Diario de Córdoba*, August 29, 1905, 2.

33. María Luz González Peña, "Rafael Bezares," *Diccionario de la zarzuela. España e Hispanoamética*, ed. Emilio Casares (Madrid: ICCMU, 2008), 1:261.

34. For example, in the "Brindis" from *Marina* (Biblioteca Nacional de España, CL/84) and the romanza from the same opera ("Costas la del Levante") (Eresbil Archivo de la

Música Vasca, FA60/199), as well as the tenor aria from *La Gioconda* (Eresbil Archivo de la Música Vasca, FA60/200), all recorded for Viuda de Aramburo.

35. Leech-Wilkinson, *The Changing Sound of Music*, chapter 4, paragraph 7 (https://charm.rhul.ac.uk/studies/chapters/chap4.html#par7). Tenors recorded particularly well too.

36. Some of Galvany's most valuable recordings (others are too deteriorated) include *Amleto*'s Rondeau (FA60/004), *L'africaine*'s "Adieu mon rivage" (FA60/005), and *La sonnambula*'s "Ah non credea mirarti" (FA60/006). They are all held at Eresbil Archivo de la Música Vasca, and were recorded by Hugens y Acosta: there is no evidence that Galvany recorded for any other *gabinete*.

37. Kelly, *The Gramophone Company Limited*; Spottswood, *Ethnic Music on Records*. Velasquez Ospina discusses how this global strategy was implemented in Colombia, in the midst of local tensions concerning class and race: Velasquez Ospina, Juan Fernando, *(Re)sounding Cities*, 250–1.

38. Bolig, *The Victor Discography. Green, Blue, and Purple Labels (1910–1926)*.

39. Bolig, *The Victor Black Label Discography*.

40. Odeón and Fonotipia, *Discos impresionados por las dos caras. Marcas Odeón y Fonotipia* (Catalogue) (Barcelona, undated).

41. Odeón and Fonotipia, *Catálogo General de Discos Odeón y Fonotipia* (Catalogue) (Barcelona, undated).

42. For example, *Dinorah*'s "Ombra leggera," which she recorded for Puerto y Novella (FA60/018 at Eresbil Archivo de la Música Vasca), privately for Regordosa (Biblioteca de Catalunya, CIL-333), and for Gramophone (053073); or *Linda di Chamounix*'s "O luce di quest'anima" for G&T (53141 7241F, in 1902–1903), Victor (52529), and privately for Regordosa (Biblioteca de Catalunya, CIL-328).

43. Eresbil Archivo de la Música Vasca FA60/074.

44. Eresbil Archivo de la Música Vasca FA60/032.

45. "Noticias de teatros," *La vanguardia*, September 10, 1895, 3; "Crónica," *La libertad*, July 20, 1897, 2; "Teatros," *El pueblo*, August 15, 1904, 3.

46. These recordings include the duet from *Lohengrin* with a señorita Vallrosoll (who is not known to have had any kind of stage career) for Centro Fonográfico Comercial de Manuel Moreno Cases (Biblioteca Nacional de España, CL/87); a passage from *Die Walküre* for Sociedad Artístico-Fonográfica; and passages from these two operas for Regordosa. All cylinders from the Biblioteca de Catalunya: "Salida de *Lohengrin*, cantada por el distinguido tenor, señor Constantí", CIL-67; "Despedida del cisne, de la ópera *Lohengrin*, cantado por el eminente tenor, señor Constantí," CIL-68; "Cant de la primavera, de *La valquíria* d'en Wagner, cantat per l-eminent artista tenor Constantí," CIL-76 and CIL-77; "Wagner, *Valquíria*, gran escena de la daga, cantada por el eminente tenor, señor Constantí," CIL-75.

47. For example in *Cavalleria rusticana*'s "Viva il vino" for Sociedad Artístico-Fonográfica, Biblioteca Nacional de España, CL/63.

48. "Lamberto Alonso," *Boletín fonográfico*, 4 (1900), 56–7.

49. "Lamberto Alonso," 57.

50. Digitized recordings include passages from *La Dolores* (FA60/185), the Epilogue from *Mefistofele* (FA60/212), and the *racconto* from *Lohengrin*, unusually split into two cylinders (FA60/214 and FA60/217). All the previously mentioned cylinders are held at Eresbil Archivo de la Música Vasca.

51. José Climent, "Lamberto Alonso," *Diccionario de la música valenciana*, ed. Emilio Casares, Vicente Galbis López, and Rafael Díaz Gómez (Valencia: Institut Valencià de la Músical, 2006), 1:41–2.

52. "Espectáculos," *Heraldo de Madrid*, November 10, 1902, 3.

53. "Noticias de Madrid," *El correo español*, May 13, 1893, 3; "En el Círculo Mercantil," *La correspondencia de España*, June 11, 1897, 3; "Concierto en el Conservatorio," *La época*, May 18, 1899, 3.

54. "Teatro de la Zarzuela," *La correspondencia militar*, March 28, 1906, 2.

55. Biblioteca Nacional de España, CL/100.

56. Biblioteca Nacional de España, CL/356.

57. "Noticias de espectáculos," *El día*, June 14, 1894, 3.

58. "Parish," *Revista contemporánea*, 109 (1898), 335.

59. A. [full name unknown], "Parish," *La ilustración española e hispanoamericana*, February 8, 1898, 83.

60. "Teatro de Parish," *La Iberia*, February 6, 1898, 3.

61. "Teatro del Buen Retiro," *El globo*, June 9, 1898, 3.

62. E. [first name unknown] Gutiérez-Gamero, "*Hänsel y Gretel*," *La ilustración española e hispanoamericana*, December 15, 1901, 339–41.

63. For example in her recording of the "Brindis" from *La viejecita* for Hugens y Acosta (Biblioteca Nacional de España, CL/300).

64. At the time when Lánderer recorded for *gabinetes* in Valencia, he was about to leave the city to train further in Italy, but, whether this happened or not, there is no evidence that he had a stage career; see "Teodoro Lánderer," *Boletín fonográfico*, 37 (1901), 21.

65. Although this could be a misspelling for Cecchini or Cechini, no performer of that name is known to have existed either.

66. T. [full name known], "La señorita Concha Sanz Arnal," *Boletín fonográfico*, 21 (1900), 344–6.

67. "Luisa Vela," *Boletín fonográfico*, 16 (1900), 263.

68. Other singers who recorded for *gabinetes* and for Gramophone during the latter's visits to Madrid and Barcelona between 1899 and 1903 (before opening a branch in Barcelona) include Rafael Bezares, Bernardino Blanquer, Manuel Figuerola, Mariano Gurrea, Concepción Carceller, Julia Mesa, Lolita Escalona, Marina Gurina, Concepción Segura, Ascensión Miralles, Felisa Lázaro, Josefa López, Ramona Galán, and Daniel Blanquells. In these early visits, however, Gramophone also recorded a few performers who are not known to have recorded for the *gabinetes*, such as Enrique Bent, Vicente Carrión, Rafael Gil, Amparo Astort, and, perhaps most interestingly, María Gay, who went on to sing at La Scala and the Met and recorded for Victor at a later stage; see Victor, *His Masters' Voice. Celebrity Records by International Artists. With Special Biographies and Photographs* (Catalogue) (no location, 1914). For the list

of performers who recorded for Gramophone and dates, see Kelly, *The Gramophone Company Limited*.

69. For example, in "Luchando tercos y rudos" from *Gigantes y cabezudos* for Viuda de Aramburo (Biblioteca Nacional de España, CL/313).

70. For example, Manuela Cubas recording "Canción de la viejecita" from *La viejecita* for Viuda de Aramburo (Eresbil Archivo de la Música Vasca, FA60/056).

71. For example, "Canción de la viejecita" from *La viejecita* for Viuda de Aramburo (Biblioteca Nacional de España, CL/312) and "Luchando tercos y rudos" for Gramophone (Biblioteca Nacional de España, DS/14055/10).

72. "Señorita Cardenal," *Boletín fonográfico*, 6 (1900), 86.

73. "Señorita Cardenal," 86.

74. "Couplets" from *Los presupuesto de Villapierde*, FA60/058, and "Romanza" from *El cabo primero*, FA60/059 (both recorded for Puerto y Novella and held at Eresbil Archivo de la Música Vasca).

75. Such as in the duet from *La leyenda del monje*, with Jesús Valiente for Puerto y Novella (FA60/269, Eresbil Archivo de la Música Vasca).

76. Indeed, there is only one book focusing on singing in *zarzuela*: Ramón Regidor Arribas, *La voz en la zarzuela* (Madrid: Real Musical, 1991). Some basic insights are provided too in Emilio Casares, "Voz," *Diccionario de la zarzuela. España e Hispanoamética*, ed. Emilio Casares (Madrid: ICCMU, 2008), 2:941–4.

77. Leech-Wilkinson, *The Changing Sound of Music*; Philip, *Performing Music in the Age of Recording*.

78. Kraft, *Stage to Studio. Musicians and the Sound Revolution*.

79. Losa, *Máchinas fallantes*, 106.

Chapter 7

1. For example anonymous, "La velocidad en la impression," *Boletín fonográfico*, 25 (1901), 11–12.

2. For example, Carl Stumpf started in 1900 an archive of field recordings, primarily ethnographic, initially at the Psychologisches Institut of the University of Berlin; see Susanne Ziegler, *Die Wachszylinder des Berliner Phonogramm-Archivs* (Berlin: Ethnologisches Museum Staatliche Museen zu Berlin, 2005). The British Museum started its collection of recordings of poets and statesmen in 1905 (now part of the British Library Sound Archive). Gram-o-phone, *Catalogue "De luxe"* (catalogue) (London, 1908) claims that several museums in Europe had started collections of recordings, but no names or further details were given.

3. "Aplicaciones del fonógrafo," *El cardo*, November 8, 1900, 20.

4. Sociedad Fonográfica Española Hugens y Acosta, *Catálogo*, 3.

5. Jaione Landaberea, "Ybarra Family Fonds." As with the other collections, the figure given here for the number of total cylinders differs slightly from those provided in the introduction to this book; in fact, the figure provided in the introduction referred

solely to commercial cylinders produced in Spain (by the *gabinetes*), and therefore excludes a very small number of foreign cylinders and of home-made recordings; these will be discussed later in this chapter.

6. Barreiro and Marro, *Primeras grabaciones fonográficas en Aragón*. Marro, who donated the cylinders to the BNE, is a descendent of Aznar.

7. *Mensaje y programa de la Cámara Agrícola del Alto Aragón* (Madrid: Imprenta de San Francisco de Sales, 1898).

8. Carrasco Mas, *Puesta en valor de la colección de cilindros de cera del Museo de Historia de la Telecomunicación Vicente Miralles Segarra*.

9. Barreiro, *Antiguas grabaciones fonográficas aragonesas*, 56.

10. The Hugens y Acosta cylinder was a chorus from the *zarzuela Los Camarones* (ES/ AHPHU - CL/000033/000007); Puerto y Novella recordings include *sevillanas* (ES/ AHPHU - CL/000033/000006) and an undigitized excerpt of *Rigoletto* (ES/AHPHU - CL/0000034/000005). The collection's catalogue, as well as digitizations of most cylinders, can be accessed through http://dara.aragon.es/opac/app/item/?vm=nv&q =hugens&p=0&i=563221.

11. Hijos de Blas Cuesta, *Hijos de Blas Cuesta* (catalogue) (Valencia, ca. 1900).

12. Sociedad Fonográfica Española Hugens y Acosta, *Sociedad Fonográfica Española Hugens y Acosta* (catalogue) (Madrid, ca. 1899).

13. Philiph Hauser, *Madrid bajo el punto de vista médico-social. Su política sanitaria, su climatología, su suelo y sus aguas, sus condiciones sanitarias, su demografía, su morbicidad y su mortalidad. 2 volúmenes (Primero y Segundo) (Edición preparada por Carmen del Moral)* (Madrid: Editora Nacional, 1979), 496.

14. Francisco Javier Fernández Roca, "El salario industrial en Sevilla: 1900–1975," *Industria y clases trabajadoras en la Sevilla del siglo XX*, ed. Carlos Arenas (Sevilla: Secretariado de Publicaciones de la Universidad de Sevilla, 1994), 115–42 (p. 120).

15. Pedro María Pérez Castroviejo, "Precios, salarios reales y estaturas en el curso de la industrialización del País Vasco, 1880–1936," *Salud y ciudades en España, 1880–1940. Condiciones ambientales, niveles de vida e intervenciones sanitarias*, ed. Francisco Muñoz, Pedro Fatjó, and Josep Pujol-Andreu (Barcelona: Universitat Autònoma de Bacelona), 8.

16. Sociedad Anónima Fonográfica, *Los reyes—Ningún regalo tan bonito y barato para este día como un fonógrafo, con 6 cilindros: 80 pesetas* (advertisement), *La correspondencia de España*, January 3, 1901.

17. Álvaro Ureña, *Álvaro Ureña* (advertisement), *ABC*, April 16, 1903, 4.

18. Cilindrique, "A los industriales y aficionados," *El cardo*, November 22, 1900, 20; "Nuestra revista," 20; Cilindrique, "Cosas fonográficas," 14.

19. Edisonia Limited, *Catalogue. Phonographs, graphophones, record supplies. Price List, 1898* (London, 1898).

20. Pathé Freres London Limited, *Pathé Frères London Limited* (catalogue) (London, 1900).

21. For example, Scottish agricultural workers earned £51 per year in 1892. Data about wages come from Arthur Lyon Bowley, *Wages in the United Kingdom in the Nineteenth Century* (Cambridge: Cambridge University Press, 1900).

22. Henri Lioret, *Description et Prix des Differents Modeles de Lioretgraph* (catalogue) (Paris, ca. 1900).

23. 140 francs per month; see François Simiand, "Le salaire des ouvriers des mines de charbon en France," *Journal de la société statistique de Paris*, 49 (1908), 13–29 (p. 23).

24. Maria Dolors Llopart, "Les cases de l'Eixample de portes endins," *La formació de l'Eixample de Barcelona: aproximacions a un fenomen urbà*, ed. Santi Barjau (Barcelona: Olimpíada Cultural, 1990), 115–27 (pp. 117–18).

25. Ospina Romero, Sergio, "Ghosts in the Machine and Other Tales," 22; Elodie A. Roy, "'You Ought to See My Phonograph': The Visual Wonder of Recorded Sound," *The Making of English Popular Culture*, ed. John Storey (Oxford and New York: Routledge, 2016), 184–97 (pp. 189–90); Nathan David Bowers, "Creating a Home Culture for the Phonograph: Women and the Rise of Sound Recordings in the United States, 1877–1913" (Ph.D. diss., University of Pittsburgh, 2007), iv. Interestingly, both Roy and Bowers focus on the role accorded to women in Britain and the United States, respectively, in facilitating the entry of the phonograph into the home (a female realm at the time). Such discourses, however, are not to be found in Spain, where phonography remained an eminently male pursuit.

26. Jorge Uría, "El nacimiento del ocio contemporáneo. Algunas reflexiones sobre el caso español," *Fiesta, juego y ocio en la historia*, ed. Ángel Vaca Lorenzo (Salamanca: Ediciones Universidad de Salamanca, 2002), 347–82.

27. Uría, "El nacimiento del ocio contemporáneo," 360.

28. Gronow, "The Record Industry," 60.

29. "Noticias," *Boletín fonográfico*, 26 (1901), 36–8.

30. Isidro Sánchez Sánchez, "Las luces del 98. Sociedades eléctricas en la España finisecular," *Sociabilidad fin de siglo. Espacios asociativos en torno a 1898*, ed. Isidro Sánchez Sánchez and Rafael Villena Espinosa (Cuenca: Ediciones de la Universidad de Castilla-La Mancha, 1999), 151–223 (p. 152).

31. Litvak, *Transformación industrial y literatura en España*, 11.

32. Don Sebastián [pseudonym], "La fiera," *El Noroeste*, April 6, 1905, 1.

33. Sophie Maisonneuve, "De la 'machine parlante' à l'auditeur le disque et la naissance d'une nouvelle culture musicale dans les années 1920–1930," *Terrain*, 37 (2001), 11–28.

34. All three collections include small numbers of foreign cylinders: Rodríguez's includes a cylinder labeled as "Banda Americana" kept inside an Hugens y Acosta box which could have been imported from abroad; the Ybarras owned a few Edison cylinders of wind band music, and Pedro Aznar similarly had cylinders (mostly opera) recorded by the Anglo-Italian Commerce Company, the Milan-based agency for Italy of Zonophone.

35. Two *tiples* by the surname Martínez, Pura and Antonietta, recorded for Gramophone in the early years of the twentieth century; there was a further *tiple*, Salud Martínez, born in Seville but active in Mexico in the late nineteenth century. It is not clear which of these recorded for Corrons.

36. For example, Fonotipia's records were signed individually by the singer, and the label encouraged customers to double-check that the signature was included as proof of authenticity and quality (Società Italiana di Fonotipia, *Catalogo* [Milan, 1907]).

37. Numbers do not include instrumental versions of specific vocal numbers.

38. These include Josefina Huguet's "Saper vorreste" for Viuda de Aramburo (FA60/029) and Puerto y Novella (FA60/012), as well as three recordings of "Sí, mi chiamano Mimí," two for Viuda de Aramburo (FA60/017 and FA60/022; these are different recordings) and one for Puerto y Novella (not catalogued).

39. "El fonógrafo," *La correspondencia de España*, November 19, 1902, 3.

40. FA60/032 at Eresbil.

41. FA60/012 at Eresbil.

42. Edisonia Limited, *Catalogue Phonographs, Graphophones, Record Supplies*.

43. Gramophone and Typewriter Ltd., Untitled catalogue (London and Sydney, 1904).

44. See for example Victor, *El libro Victrola de la ópera* (Camden, NJ: Victor Talking Machine Company, 1925), in which synopses of the operas considered the most significant were printed together with suggestions of recordings from Victor's catalogue. Such publications disseminated a notion of high culture throughout the world; see Velásquez Ospina, Juan Fernando, *(Re)sounding Cities*, 272–3.

45. Don Sebastián, "La Fiera."

46. Jacinto Benavente, "Noches de verano," *La correspondencia de Alicante*, July 21, 1897, 1.

47. "Grata reunión," *Diario de Córdoba*, August 22, 1903, 2.

48. "Ecos," *Diario oficial de avisos de Madrid*, May 2, 1901, 2.

49. "La velocidad en la impresión," and "Membranas y bocinas," *Boletín fonográfico*, 25 (1901), 11–12 and 13–14.

50. Marín, "La impresión de fonogramas," 19–20.

51. "Concurso fonográfico," *Boletín fonográfico*, 1 (1900), 11.

52. "Correspondencia," *Boletín fonográfico*, 11 (1900), 146.

53. CL/205.

54. CL/382.

55. CL/308 at the Biblioteca Nacional de España (same for the following notes).

56. CL/324.

57. CL/378, CL/395.

58. Works recorded included the following: Saint-Saëns's *Rondeau capricieux* (split over two cylinders), Pablo Sarasate's *Capricho vasco* (split over two cylinders), and *Jota*, Bazzini's *Ronde des lutins*, a fantaisie on *Lucia di Lammermoor*'s concertante, and an arrangement of Chopin's Nocturne, op. 9 no. 2.

59. Barreiro, *Antiguas grabaciones fonográficas aragonesas*, 8 and 56.

60. Wenceslao Retana, "Unos días en Huesca (Impresiones de un ex gobernador)," *Por esos mundos*, 155 (1907), 514–26.

61. Cited in Barreiro, *Antiguas grabaciones fonográficas aragonesas, 1898–1907*, 64.

62. *The Dawn of Recording: The Julius Block Cylinders*, Marston Records, 3 CDS (West Chester, 2008). Block's collection, including spoken word cylinders, amounts to ca. 215 recordings.

63. New York Public Library, *The Mapleson Cylinders*, New York Public Library, 6 LPs (New York, 1985).

64. The breakdown by genre is as follows: 127 recordings of opera, 29 of *género chico*, 24 of art song (mostly Spanish and Catalan), 15 of flamenco, 12 of Neapolitan song, 11 of traditional music, 9 of cabaret, 7 of choral music, 3 each of operetta and *zarzuela grande*, and 1 of oratorio.

65. "Notas locales," *La vanguardia*, January 28, 1896, 2; "Notas locales," *La vanguardia*, July 27, 1902, 2; "Notas locales," *La vanguardia*, January 6, 1904, 2; "Notas locales," *La vanguardia*, January 27, 1906, 2; "Notas locales," *La vanguardia*, May 21, 1908, 2.

66. "Notas locales," *La vanguardia*, April 3, 1912, 2.

67. "Espectáculos," *La vanguardia*, May 4, 1903, 4.

68. "Racconto di Santuzza en la Cavalleria, por la distinguida artista Sra. Gurina," Biblioteca de Catalunya, CIL-207.

69. "Romanza de La Trapera, por la distinguida artista Srta. Gurina," Biblioteca de Catalunya, CIL-202; "Romanzas de Gigantes y cabezudos, por Marina Gurina," Biblioteca de Catalunya, CIL-215.

70. "Aria de la Doloretes, por la distinguida artista Srta. Gurina," Biblioteca de Catalunya, CIL-216.

71. "Serenata de Schubert, cantada por la eminente diva Josefina Huguet," Biblioteca de Catalunya, CIL-346.

72. Julia Culp (perf.), Franz Schubert (comp.), Serenade (Camden, NJ, 1914), 74431.

73. Gelatt, *The Fabulous Phonograph*, 103.

74. Edison Bell Phonographs, *Edison Bell Phonographs. British Home of the Phonograph* (catalogue) (London, 1905); Brown Brothers Limited, *Phonographs. Columbia Graphophones. Horns. Phonograph Parts & CO* (catalogue) (London, 1906); Indestructible Records, *Indestructible Records* (catalogue) (unknown location, 1908).

75. Read and Welch, *From Tin Foil to Stereo*, 107.

76. Gelatt, *The Fabulous Phonograph*, 104.

77. Gelatt, *The Fabulous Phonograph*, 168.

78. "Ecos", 2.

Conclusion: From national to global

1. Gronow, "The Record Industry", 56.

2. Examples can be found in: Pekka Gronow and Björn Englund, "Inventing Recorded Music", 292; Losa, *Máchinas fallantes*, 172.

3. Suisman, *Selling Sounds*; Denning, *Noise Uprising*.

4. Gitelman, *Scripts, Grooves, and Writing Machines*; Sterne, *The Audible Past*.

5. Gronow and Englund also mention, albeit in passing, similar theatrical repertoires in Scandinavia: Gronow and Englund, "Inventing Recorded Music", 291.

Bibliography

Primary sources

La corona, 8 April 1863, p. 2.
El papa-moscas, 74 (1879), p. 2.
La publicidad, 4 April 1879, p. 3.
Diari catalá, 25 June 1879, p. 3.
Diari catalá, 24 July 1879, p. 3.
Diari catalá, 21 September 1879, p. 3.
Diari catalá, 28 September 1880, p. 3.
Lo gay saber, 1 November 1880, p. 1.
Lo gay saber, 10 November 1880, p. 11.
La correspondencia de España, 10 February 1882, p. 2.
La correspondencia de España, 19 August 1882, p. 3.
La discusión, 20 August 1882, p. 3.
Crónica de la música, 23 August 1882, p. 1.
El liberal, 29 August 1882, p. 5.
El imparcial, 14 September 1882, p. 2.
El correo militar, 29 November 1883, p. 3.
El liberal, 17 April 1885, p. 3.
La correspondencia de España, 20 June 1885, p. 2.
La unión, 29 August 1885, p. 3.
La Iberia, 31 August 1885, p. 3.
El nuevo Ateneo, 15 March 1887, p. 5.
El constitucional, 16 December 1887, p. 1.
El correo militar, 26 March 1890, p. 3.
La vanguardia, 27 December 1892, p. 3.
El diario de Córdoba, 21 March 1894, p. 2.
El reservista, 19 January 1895, p. 3.
Diario oficial de avisos de Madrid, 6 December 1895, p. 2.
El bien público, 7 December 1895.
Crónica meridional, 31 January 1896, p. 3.
El correo militar, 9 December 1896, p. 3.
El Ateneo, 10 July 1898, p. 7.
Industria e invenciones, 22 April 1899, pp. 155–6.
La publicidad, 23 April 1899, p. 3.
La música ilustrada hispano-americana, 25 April 1899, p. 10.
La correspondencia de España, 2 January 1900, p. 3.
La época, 15 February 1900, p. 2.
La correspondencia militar, 12 October 1900, p. 2.
El cardo, 15 February 1901, p. 14.
El cardo, 22 February 1901, p. 15.

El cardo, 8 March 1901, p. 15.

Heraldo de Alcoy, 30 April 1901, p. 3.

La época, 15 May 1902, p. 2.

ABC, 19 August 1911, p. 12.

"El fonógrafo", *El constitucional*, 23 January 1878, p. 1.

"Sección de noticias", *El progreso de Lugo*, 26 January 1878, p. 3.

"Miscelánea", *El Ateneo*, 21 March 1878, p. 8.

"Semana histórica", *La academia*, 23 March 1878, p. 2.

"Ateneo Barcelonés", *La publicidad*, 9 September 1878, p. 1.

"Noticias generales", *La época*, 12 January 1879, p. 4.

"Revista de la semana", *La mañana*, 19 January 1879, p. 1.

"Edición de la noche", *La correspondencia de España*, 6 February 1879, p. 2.

"Semana histórica", *La academia*, 15 February 1879, p. 3.

"Cridas", *Lo nunci*, 20 April 1879, p. 4.

"Noticias locales y generales", *El comercio*, 10 June 1879, p. 2.

"Novedades teatrales", *El globo*, 4 January 1880, pp. 3–4.

"Sesiones", *Diario de Córdoba*, 20 January 1880, p. 1.

"Fonógrafo Edison", *Diario de Córdoba*, 12 February 1880, p. 2.

"Descargas cerradas", *El Fígaro*, 14 February 1880, p. 1.

"Gacetillas", *El graduador*, 19 February 1880, p. 2.

"Gacetillas", *El graduador*, 24 February 1880, p. 2.

"Crónica", *El eco de Cartagena*, 28 February 1880, p. 2.

"Liceo", *Diario de Murcia*, 3 March 1880, p. 1.

"Popularisasió del teléfono", *Diari catalá*, 24 September 1880, p. 3.

"Conferencias en Círculo de la Unión Católica", *El magisterio español*, 20 January 1882, p. 2.

"Jochs florals", *La ilustració catalana*, 1 February 1882, p. 13.

"Diversiones públicas", *El liberal*, 14 February 1882, p. 3.

"Teatro de la Comedia: exhibición de fonógrafo por el Dr. Llop [sic]", *El globo*, 14 February 1882, p. 3.

"El fonógrafo", *El día*, 19 August 1882, p. 3.

"Edición de la tarde", *La correspondencia de España*, 20 September 1882, p. 2.

Exposición Universal de Barcelona: año 1888. Clasificación de productos (Barcelona: publisher unknown, 1887).

Exposición universal de Barcelona. Catálogo oficial especial de España (Barcelona: Imprenta de los Sucesores de N. Ramírez, 1888).

"Noticias", *La correspondencia de España*, 1 December 1892, p. 3.

"Jochs florals", *La ilustració catalana*, 30 January 1883, p. 14.

"Jardín del buen Retiro", *El popular*, 22 August 1884, p. 3.

"Sección de espectáculos", *El busilis*, 12 December 1884, p. 4.

"En Recoletos: *El fonógrafo*, obra de teatro de Castillo y Soriano", *El imparcial*, 30 August 1885, p. 3.

"*El fonógrafo*", *La correspondencia de España*, 9 September 1885, p. 2.

"Gacetillas", *Crónica de Cataluña*, 7 April 1886, p. 1.

"Provincias", *La justicia*, 23 August 1888, p. 3.

"El fonógrafo y el grafófono", *La gaceta industrial*, 25 December 1888, p. 9.

"Una feria fantástica", *Diario de Tenerife*, 29 January 1890, p. 4.

"Espectáculos", *La publicidad*, 2 February 1890, p. 3.

"Teatro Circo", *La vanguardia*, 28 February 1890, p. 2.

"Espectáculos", *La publicidad*, 2 March 1890, p. 3.

"Noticias y avisos", *La nueva lucha: diario de Gerona*, 4 March 1890, p. 3.

"Notas locales", *La vanguardia*, 5 March 1890, p. 7.

"Teatro de San Andrés de Palomar", *La vanguardia*, 6 September 1890, p. 2.

"Gacetilla", *El bien público*, 28 June 1892, p. 2.

"Gacetilla", *El liberal*, 2 July 1892, p. 3.

"Fonógrafo", *Diario de Córdoba*, 20 August 1892, p. 3.

"El fonógrafo perfeccionado de Edisson [sic]", *La correspondencia de España*, 2 November 1892, p. 3.

"Diversiones públicas", *La época*, 4 November 1892, p. 3.

Recopilación de estudios e investigaciones efectuada por la Comisión Obrera Catalana en la Exposición de Chicago (Barcelona: unknown publisher, 1893).

"Noticias de Madrid", *El correo español*, 13 May 1893, p. 3.

"Sucesos", *La lealtad navarra*, 6 July 1893, p. 2.

"Crónica", *El tradicionalista*, 6 July 1893, p. 6.

"Gacetillas", *El heraldo de Madrid*, 11 January 1894, p. 4.

"Noticias", *La atalaya*, 28 January 1894, p. 3.

"Fonda de Oriente", *Diario de Córdoba*, 21 March 1894, p. 3.

"El fonógrafo Edison", *Diario de Córdoba*, 30 March 1894, p. 3.

"Espectáculos", 17 April 1894, p. 4.

"Notable fonógrafo", *El popular*, 23 May 1894, p. 2.

"Noticias de espectáculos", *El día*, 14 June 1894, p. 3.

"Sección de noticias", *La Rioja*, 17 June 1894, p. 2.

"El fonógrafo en el Teatro", *La Rioja*, 24 June 1894, p. 2.

"Espectáculo científico", *La correspondencia de España*, 27 June 1894, p. 4.

"La Rioja en Haro", *La Rioja*, 5 September 1894, p. 1.

"Sección de noticias", *La Rioja*, 19 September 1894, p. 2.

"Fonógrafo", *La región extremeña*, 23 October 1894, p. 2.

"El fonógrafo", *Crónica meridional*, 22 December 1894, p. 2.

"Crónica general", *La publicidad*, 8 January 1895, p. 2.

"Fonógrafo", *La Iberia*, 10 January 1895, p. 3.

"Noticias", *La correspondencia de España*, 3 February 1895, p. 2.

"Madrid", *El correo militar*, 11 February 1895, p. 2.

"Gacetillas", *El Guadalete*, 5 March 1895, p. 3.

"Noticias", *El correo militar*, 9 March 1895, p. 2.

"Edición de la tarde", *Heraldo de Baleares*, 2 April 1895, p. 3.

"Madrid", *La correspondencia de España*, 28 April 1895, p. 2.

"El kinetóscopo de Edisson [sic]", *El día*, 14 May 1895, p. 3.

"Gacetillas", *El Guadalete*, 16 March 1895, p. 2.

"Noticias", *El día*, 18 March 1895, p. 3.

"Gacetillas", *El eco de Navarra*, 7 July 1895, p. 2.

"Noticias de espectáculos", *La unión católica*, 8 July 1895, p. 3.

"Crónica general", *La dinastía*, 7 September 1895, p. 2.

"Noticias de teatros", *La vanguardia*, 10 September 1895, p. 3.

"Ecos de Sóller", *Heraldo de Baleares*, 5 August 1895, p. 2.

"Crónica local", *Heraldo de Baleares*, 19 October 1895, p. 2.

"El fonógrafo testigo", *La lucha*, 8 November 1895, p. 3.

"Fonógrafo Edisson [sic]", *El Thader*, 14 November 1895, p. 2.

"El fonógrafo de Segovia", *La legalidad*, 7 December 1895, p. 3.

"El Ateneo", *El país*, 15 December 1895, p. 3.

"Ateneo de Madrid", *El globo*, 16 December 1895, p. 3.

"Ateneo de Madrid", *El globo*, 16 December 1895, p. 3.

"Provincias", *El correo de España*, 5 January 1896, p. 2.

"Fonógrafo", *La atalaya*, 6 January 1896, p. 3.

"Noticias", *El guasón*, 12 January 1896, p. 3.

"Noticias", *La lucha*, 14 January 1896, p. 3.

"Notas locales", *La vanguardia*, 28 January 1896, p. 2.

"Gacetillas", *Crónica meridional*, 31 January 1896, p. 3.

"Noticias", *El baluarte*, 23 February 1896, p. 3.

"Gacetillas", *El Guadalete*, 7 March 1896, p. 2.

"Menestra", *La Rioja*, 19 April 1896, p. 2.

"Gacetillas", *El vigía católico*, 27 June 1896, p. 3.

"Noticias", *El carpetano*, 2 July 1896, p. 4.

"Leemos en *La Monarquía*", *El nuevo alicantino*, 8 July 1896, p. 3.

"Un fonógrafo", *El Ateneo*, 10 July 1896, p. 7.

"Crónica", *El liberal navarro*, 17 July 1896, p. 2.

"El mejor fotógrafo", *El Ateneo*, 10 August 1896, p. 8.

"Fonógrafo", *El cantábrico*, 29 August 1896, p. 3.

"Fonógrafo", *La tempestad*, 6 September 1896, p. 3.

"Espectáculos", *La vanguardia*, 29 September 1896, p. 7.

"Accésit", *El noticiero de Soria*, 3 October 1896, p. 2.

"Gacetillas", *El bien público*, 7 October 1896, p. 3.

"Crónica local", *El isleño*, 16 October 1896, p. 2.

"Pórtico de Apolo", *La correspondencia de España*, 19 October 1896, p. 3.

"Gacetillas", *El eco de Navarra*, 20 November 1896, p. 2.

"Fonógrafo", *La semana católica de Salamanca*, 5 December 1896, p. 15.

"Provincias", *La correspondencia de España*, 14 December 1896, p. 4.

"Notas de electricidad", *Madrid científico*, 146 (1897), p. 10.

"Café Colón", *La publicidad*, 27 January 1897, p. 4.

"Universal Exprés", *La vanguardia*, 14 February 1897, p. 7.

"Exposición artística a beneficio de los heridos de Cuba y Filipinas", *La época*, 24 February 1897, p. 3.

"En el Ateneo", *Diario del comercio*, 14 March 1897, p. 2.

"Madrid", *El globo*, 15 March 1897, p. 2.

"Noticias", *La voz de la provincia*, 8 April 1897, p. 2.

"Sociedad de Fomento", *Boletín republicano de la provincia de Gerona*, 24 April 1897, p. 3.

"Eliseo Exprés", 7 May 1897, *La correspondencia alicantina*, p. 2.

"Teatro Principal", *El graduador*, 18 May 1897, p. 3.

"En el Círculo Mercantil", *La correspondencia de España*, 11 June 1897, p. 3.

"Asilo de Santa Cristina", *Diario oficial de avisos de Madrid*, 4 July 1897, p. 3.

"Crónica general", *Crónica reusense*, 10 July 1897, p. 1.

"Crónica general", *Crónica reusense*, 12 July 1897, p. 1.

"Gacetillas", *El eco de Navarra*, 18 July 1897, p. 2.

"Nuevo establecimiento", *El correo militar*, 19 July 1897, p. 2.

"El correo militar", *Nuevo establecimiento*, 19 July 1897, p. 2.

"Crónica", *La libertad*, 20 July 1897, p. 2.

"Noticias locales", *El regional*, 28 July 1897, p. 2.

"Desde la corte", *El isleño*, 2 August 1897, p. 1.

"Noticias", *La opinión*, 22 September 1897, p. 2.

"Avisos y noticias", *La Rioja*, 6 October 1897, p. 2.

"Salón Express. Pórticos de Berdejo", *La voz de la provincia*, 27 November 1897, p. 3.

"Noticias", *La correspondencia de España*, 28 November 1897, p. 3.

"La velada del Círculo Mercantil", *La correspondencia de España*, 19 December 1897, p. 1.

"Círculo de la Unión Mercantil", *La época*, 19 December 1897, p. 2.

Guía práctica de Valencia (Valencia, 1898: Imprenta de José Ortega).

Mensaje y programa de la Cámara Agrícola del Alto Aragón (Madrid: Imprenta de San Francisco de Sales, 1898).

"Parish", *Revista contemporánea*, 109 (1898), p. 335.

"Gacetillas", *Diario de Córdoba*, 2 February 1898, p. 2.

"Teatro de Parish", *La Iberia*, 6 February 1898, p. 3.

"Fonógrafo y cinematógrafo", *La vanguardia*, 4 March 1898, p. 2.

"Noticias", *La correspondencia de Alicante*, 4 May 1898, p. 2.

"Audiencia provincial", *Flores y abejas*, 22 May 1898, p. 6.

"Teatro del Buen Retiro", *El globo*, 9 June 1898, p. 3.

"Noticias", *Heraldo de Navarra*, 13 July 1898.

"Fonógrafo Edison", *El graduador*, 20 July 1898, p. 20.

"Gacetillas", *El eco de Navarra*, 5 August 1898, p. 2.

"Noticias", *La Rioja*, 25 September 1898, p. 2.

"Anuncios", *El eco de Santiago*, 28 October 1898, p. 2.

"Salón Exprés", *El noticiero de Soria*, 29 October 1898, p. 2.

"Salón Exprés", *El noticiero de Soria*, 2 November 1898, p. 2.

"Enrique Sepúlveda", *La época*, 29 December 1898, p. 1.

"Círculos y sociedades", *Las provincias de Valencia*, 22 February 1899, p. 3.

"Valencia", *Las provincias*, 1 April 1899, p. 2.

"Hojas sueltas", *Los deportes*, 15 April 1899, p. 142.

"Teatros", *La publicidad*, 21 April 1899, p. 4.

"Crónica", *La hormiga de oro*, 22 April 1899, p. 15.

"Crónica de espectáculos", *El día de Palencia*, 11 May 1899, p. 3.

"El fonógrafo ambulante", *La dinastía*, 15 May 1899, p. 1.

"Concierto en el Conservatorio", *La época*, 18 May 1899, p. 3.

"La Feria", *El día de Palencia*, 20 May 1899, p. 2.

"Ecos madrileños", *La época*, 6 July 1899, p. 2.

"Notas locales", *La vanguardia*, 21 July 1899, p. 2.

"Gacetilla", *El vigía de Ciudadela*, 22 July 1899, p. 2.

"Miscelánea", *Mallorca*, 25 July 1899, p. 432.

"Automóviles en Madrid", *La época*, 21 September 1899, p. 3.

"Madrid Artístico. Sociedad Fonográfica Española – Hugens y Acosta", *El liberal*, 3 October 1899, p. 2.

"Plaza del Teatro", *El liberal*, 9 November 1899, p. 3.

"Círculo militar", *La correspondencia de España*, 19 November 1899, p. 3.

"El fonógrafo en Valencia", *Las provincias*, 2 December 1899, p. 2.

"Los maestros", *Diario de Tortosa*, 20 December 1899, p. 2.

"A los lectores", *Boletín fonográfico*, 1 (1900), p. 3.

"Diafragma The Keating. Reproductor para gramófonos", *Boletín fonográfico*, 1 (1900), pp. 6–7.

"Concurso fonográfico", *Boletín fonográfico*, 1 (1900), p. 11.

"Manuel Pallás", *Boletín fonográfico*, 2 (1900), p. 22.

"Maximilano Thous", *Boletín fonográfico*, 3 (1900), p. 33.

"Los aranceles y el fonógrafo", *Boletín fonográfico*, 3 (1900), p. 36.

"Una real orden", *Boletín fonográfico*, 4 (1900), pp. 51–2.

"Lamberto Alonso", *Boletín fonográfico*, 4 (1900), pp. 56–7.

"Impresiones", *Boletín fonográfico*, 4 (1900), pp. 52–3.

"El maestro Bellver", *Boletín fonográfico*, 6 (1900), p. 81.

"Señorita Cardenal", *Boletín fonográfico*, 6 (1900), p. 86.

"Sres. Fiol y Villar", *Boletín fonográfico*, 6 (1900), p. 87.

"El diafragma Pallás", *Boletín fonográfico*, 6 (1900), p. 90.

"Gabinetes fonográficos españoles. II. Sres. Hijos de Blas Cuesta, de Valencia", *Boletín fonográfico*, 9 (1900), pp. 140–2.

"Por lo flamenco", *Boletín fonográfico*, 10 (1900), pp. 156–8.

"Correspondencia", *Boletín fonográfico*, 11 (1900), p. 146.

"Jesús Valiente", *Boletín fonográfico*, 12 (1900), p. 189.

"El maestro Rodríguez", *Boletín fonográfico*, 13 (1900), p. 204.

"El mercado fonográfico", *Boletín fonográfico*, 14 (1900), p. 214.

"Gabinetes fonográficos españoles. Sres. Puerto y Novella, de Valencia", *Boletín fonográfico*, 14 (1900), pp. 218–22.

"Catálogo general Puerto y Novella", *Boletín fonográfico*, 14 (1900), pp. 223–9.

"José Bayarri", *Boletín fonográfico*, 15 (1900), p. 248.

"Luisa Vela", *Boletín fonográfico*, 16 (1900), p. 263.

"Noticias", *Boletín fonográfico*, 16 (1900), p. 265.

"Gabinetes fonográficos españoles. Sres. Pallás y Comp., de Valencia", *Boletín fonográfico*, 18 (1900), pp. 294–7.

"Noticias", *Boletín fonográfico*, 19 (1900), p. 297.

"El maestro Goñi", *Boletín fonográfico*, 22 (1900), p. 364.

"Gabinete fonográfico de los señores Hércules Hermanos. Salón de impresionar", *Boletín fonográfico*, 23 (1900), p. 373.

"Gabinetes fonográficos españoles. VI. Sres. Hércules Hermanos de Valencia", *Boletín fonográfico*, 23 (1900), pp. 383–5.

"Gran subasta", *El avisador numantino*, 4 January 1900, p. 4.

"Salón Exprés", *Heraldo de Alcoy*, 11 March 1900, p. 3.

"Anuncios de interés", *El Guadalete*, 17 March 1900, p. 3.

"Noticias", *El eco de la fusión*, 22 March 1900, p. 3.

"Crónica local y general", *Diario de Reus*, 27 March 1900, p. 2.

"Audiciones fonográficas", *El país*, 17 April 1900, p. 4.

"Santiago", *El eco de Santiago*, 20 October 1900, p. 2.

"Aplicaciones del fonógrafo", *El cardo*, 8 November 1900, p. 20.

"Aquí y fuera de aquí", *El cardo*, 15 November 1900, p. 20.

"Fonogramas recibidos", *El cardo*, 30 November 1900, p. 1.

"Nuestra revista", *El cardo*, 30 November 1900, p. 20.

"Concurso fonográfico", *El cardo*, 15 December 1900, p. 14.

"El mercado fonográfico", *Boletín fonográfico*, 23 (1901), p. 382.

"La velocidad en la impresión", *Boletín fonográfico*, 25 (1901), pp. 11–12.

"Membranas y bocinas", *Boletín fonográfico*, 25 (1901), pp. 11–12 and 13–14.

"De re fonográfica", *Boletín fonográfico*, 25 (1901), p. 14.

"Francisca Segura", *Boletín fonográfico*, 26 (1901), p. 36.

"Noticias", *Boletín fonográfico*, 26 (1901), pp. 36–8.

"De re fonográfica", *Boletín fonográfico*, 28 (1901), pp. 69–70.

"De re fonográfica", *Boletín fonográfico*, 29 (1901), p. 84.

"Noticias", *Boletín fonográfico*, 29 (1901), p. 85.

"Impresión de cilindros", *Boletín fonográfico*, 32 (1901), pp. 134–5.

"Gabinetes fonográficos españoles. D. José Navarro, de Madrid", *Boletín fonográfico*, 34 (1901), pp. 167–9.

"Gabinete fonográfico", *Boletín fonográfico*, 35–6 (1901), p. 133.

"De re fonográfica", *Boletín fonográfico*, 35–6 (1901), pp. 185–6.

"Teodoro Lánderer", *Boletín fonográfico*, 37 (1901), p. 21.

"Industria fonográfica", *El cardo*, 22 January 1901, pp. 14–15.

"El fonógrafo y el gramófono", *El cardo*, 8 February 1901, p. 14.

"Industria fonográfica", *El cardo*, 22 January 1901, p. 14.

"Novedades", *El cardo*, 8 March 1901, p. 15.

"Sobre la originalidad de los cilindros", *El cardo*, 30 March 1901, p. 14.

"Ecos", *Diario oficial de avisos de Madrid*, 2 May 1901, p. 2.

"Exposición madrileña de pequeñas industrias", *El siglo futuro*, 10 June 1901, p. 1.

Untitled drawing, *El cardo*, 8 August 1901, p. 13.

"Fonógrafos", *Diario de Córdoba*, 21 November 1901, p. 4.

"Notas locales", *La vanguardia*, 27 July 1902, p. 2.

"Espectáculos", *Heraldo de Madrid*, 10 November 1902, p. 3.

"El fonógrafo", *La correspondencia de España*, 19 November 1902, p. 3.

Programa de festejos. Gran feria en la ciudad de Valencia (Valencia: no publisher, 1903).

Gran feria de Valencia. 20 al 31 de junio de 1903 (programa de fiestas) (festival booklet) (Valencia: no publisher, 1903).

"Espectáculos", *La vanguardia*, 4 May 1903, p. 4.

"Grata reunión", *Diario de Córdoba*, 22 August 1903, p. 2.

"Crónica general", *El adelanto de Salamanca*, 24 August 1903, p. 1.

"Sueltos", *La opinión*, 30 August 1903, p. 2.

"Notas locales", *La vanguardia*, 6 January 1904, p. 2.

"Teatros", *El pueblo*, 15 August 1904, 3.

"Noticias de provincias", *El globo*, 30 April 1905, p. 2.

"De Ciudadela", *El grano de arena*, 7 June 1905, p. 3.

"Nuestros artistas", *Diario de Córdoba*, 29 August 1905, p. 2.

"Notas locales", *La vanguardia*, 27 January 1906, p. 2.

"Teatro de la Zarzuela", *La correspondencia militar*, 28 March 1906, p. 2.

"Notas locales", *La vanguardia*, 21 May 1908, p. 2.

"Notas locales", *La vanguardia*, 3 April 1912, p. 2.

A. [full name unknown], "Parish", *La ilustración española e hispanoamericana*, 8 February 1898, p. 83.

Alcañiz, José, *José Alcañiz* (Advertisement), *Boletín fonográfico*, 3 (1900), p. 7.

Alcover, José, "El teléfono Bell y el teléfono de repetición de M. Edison", *La gaceta industrial*, 10 December 1877, pp. 2–4.

B., J. [full name unknown], "Carmelo Bueso Beltrán", *Boletín fonográfico*, 8 (1900), p. 120.

Baeza y Salvador, Armando, *Ciencia popular: los misterios de la ciencia* (Barcelona: Ramón Molinas, 1890).

Bargeon de Viverols, Jean-Marie, *Teatro de Apolo. Funciones para hoy martes 24 de junio de 1879* (advertisement) (24 June 1879), http://bdh-rd.bne.es/viewer.vm?id=0000032906, accessed May 2018.

Bargeon de Viverols, Jean-Marie, *Única representación del fonógrafo Edison. El invento más admirable del siglo* (advertisement) (1879), clip from an unknown publication held at the Biblioteca de Catalunya, reference number Cat 1/7.

Bargeon de Viverols, Jean-Marie, telegram to Thomas Alva Edison, 11 October 1888, The Thomas Edison Papers, Rutgers University, D8849, http://edison.rutgers.edu/NamesSearch/DocImage.php?DocId=D8849ACK&.

Benavente, Jacinto, "Noches de verano", *La correspondencia de Alicante*, 21 July 1897, p. 1.

Blasco, R. [first name unknown], "*El fonógrafo ambulante*", *La correspondencia de España*, 25 April 1899, p. 3.

Blasco Ibáñez, Vicente, "Viñas en el fonógrafo", *Boletín fonográfico*, 17 (1900), pp. 279–82.

Bowley, Arthur Lyon, *Wages in the United Kingdom in the Nineteenth Century* (Cambridge: Cambridge University Press, 1900).

Brown Brothers Limited, *Phonographs. Columbia Graphophones. Horns. Phonograph Parts & CO* (catalogue) (London, 1906).

Brugués y Escuder, Casimiro, *La cera de abejas: procedimientos para ensayar la cera de abejas y su aplicación al examen de diferentes muestras de ceras españolas procedentes de apicultores, de farmacias, de rito, de droguerías y del comercio de cerería* (Barcelona: Imprenta de Pedro Ortega, 1900).

C.[first name incomplete], "Industrias fonográficas", *Boletín fonográfico*, 33 (1901), pp. 148–9.

C., I. [full name unknown], "Desde Béjar", *El adelanto*, 8 January 1906, p. 1.

Casa Cabedo, *Novedades musicales para impresionar cilindros* (Advertisement), *Boletín fonográfico*, 9 (1900), p. 145.

Casa Cabedo, *Graduador de diafragma* (advertisement), *Boletín fonográfico*, 9 (1900), p. 131.

Casas y Barbosa, José, "Descripción del teléfono, el micrófono y el fonógrafo", *Las maravillas de la naturaleza, de la ciencia y del arte*, 5th ed., ed. Manuel Aranda y Sanjuán (Barcelona: Trilla y Serrra, no year).

Cilindrique [pseudonym], "Fonografía", *El cardo*, 8 November 1900, p. 17.

Cilindrique, "Cancanes fonográficos", *El cardo*, 15 November 1900, p. 22.

Cilindrique, "A los industriales y aficionados", *El cardo*, 22 November 1900, p. 20.

Cilindrique, "Para todos, sabios e ignorantes", *El cardo*, 15 December 1900, p. 14.

Cilindrique, "El arte y el gramófono", *El cardo*, 22 December 1900, p. 14.

Cilindrique, "Fonografía madrileña", *El cardo*, 15 April 1901, p. 14.

Cilindrique, "Cosas de fonografía", *El cardo*, 30 March 1901, pp. 15–16.

Cilindrique, "Cosas fonográficas", *El cardo*, 8 June 1901, p. 14.

Cilindrique, "Cosas fonográficas", *El cardo*, 22 June 1901, p. 14.

Cilindrique, "De fonografía", *El cardo*, 8 September 1901, p. 14.

Cilindrique, "Sección fonográfica", *El cardo*, 8 October 1901, p. 14.

Cilindrique, "Asuntos fonográficos", *El cardo*, 8 October 1901, p. 14.

Cilindrique, "Cancanes fonográficos", *El cardo*, 15 November 1900, p. 22.

José Corrons, *Óptica Corrons* (advertisement), *La vanguardia*, 13 December 1900, p. 6.

Corrons, José, *Canals & Corrons* (advertisement), *Boletín fonográfico*, 35–6 (1901), p. 194.

Cuesta y Sáinz, Antonio de la, "Cuento", *El correo ibérico*, 30 December 1904, p. 2.

Don Sebastián [pseudonym], "La fiera", *El Noroeste*, 6 April 1905, p. 1.

Edison, Thomas A., "The Perfected Phonograph", *The North American Review*, 146, no. 379 (1888), pp. 641–50.

Edison, Thomas A., "El nuevo fonógrafo de Edisson [sic]", *La publicidad*, 2 August 1888, p. 2.

Edison, Thomas A., "El fonógrafo perfeccionado", *Crónica científica*, 25 September 1888, pp. 411–14.

Edison, Thomas A., *Description of the Phonograph and Phonograph-Graphophone by Their Respective Inventors. Testimonials as to Their Practical Use* (Edison: New York, 1888).

Edison Bell Phonographs, *Edison Bell Phonographs. British Home of the Phonograph* (catalogue) (London, 1905).

Edison Bell, *Up to Date Dealer's Special Parcel* (Catalogue) (unknown location, 1906).

Edisonia Limited, *Phonographs, Graphophones, Record Supplies. Price List* (Catalogue) (London, 1898).

El graphos, *El graphos* (advertisement), *La region extremeña*, 23 March 1900, p. 4.

Estradé Simó, Juan Bautista, and Ricardo Ribas Anguera, *Ribas y Estradé* (advertisement), *La vanguardia*, 31 October 1899, p. 3.

Estremera, José, "Microscopio gigante", *Madrid cómico*, 31 August 1884, p. 2.

Folch y Torres, José María, *Barcelona: itinerarios para visitar la ciudad, Exposición Internacional de Industrias Eléctricas y Nacional de Productos: guía práctica para el turismo* (Barcelona: Thomas, 1914).

Fonógrafo Soley, *Fonógrafo Soley* (advertisement), *La publicidad*, 30 August 1893, p. 7.

Fono-Reyna, *Fono-Reyna* (advertisement), *La correspondencia de España*, 12 August 1903, p. 4.

G. M., A. [full name unknown], "El fonógrafo", *La semana católica de Salamanca*, 5 June 1886, p. 10.

Garci-Fernández, [first name unknown], "Política europea", *Crónica meridional*, 26 January 1900, p. 1.

Garci-Fernández, [first name unknown], "Política europea", *La región extremeña*, 8 February 1900, p. 1.

Garci-Fernández, [first name unknown], "Política europea", *La región extremeña*, 1 March 1900, p. 1.

Girona, Manuel, *Memoria reglamentaria de la Exposición Universal de Barcelona de 1888* (Barcelona: Heinrich, 1889).

Gómez R., M. [first name unknown], *Guía comercial de Valencia, Alicante y Castellón de la Plana* (Valencia: no publisher, 1904).

Grammophon, *Grammophon. Verzeichnis französischer italienischer spanischer portugiesischer Platten* (catalogue), no date (estimated ca. 1900) (Biblioteca Nacional de Catalunya, Cat 1/2).

Gram-o-phone, *Catalogue "De luxe"* (catalogue) (London, 1908).

Gramophone, *El Gramophone* (advertisement), *La atalaya*, 5 August 1903, p. 4.

Gramophone and Typewriter Ltd., Untitled catalogue (London and Sydney, 1904).

Gramophone, *Gramophone noticias* (advertisement), *El porvenir segoviano*, 24 February 1904, p. 4.

Gramophone, *Libro de oro de la Compañía Francesa del Gramófono* (Barcelona: Gramophone, 1904).

Gramophone/His Master's Voice, *Orfeó Catalá* (catalogue), 1916 (Biblioteca Nacional de Catalunya, Cat 8/11).

Grandes Almacenes El Siglo, *Grandes Almacenes El Siglo* (advertisement), *La vanguardia*, 16 February 1899, p. 7.

Gresca, Armando, "Crónica teatral", *El arte del teatro*, 28 (1907), p. 1.

Gutiérez-Gamero, E.[first name unknown], *"Hänsel y Gretel"*, *La ilustración española e hispanoamericana*, 15 December 1901, pp. 339–41.

Hauser, Philiph, *Madrid bajo el punto de vista médico-social. Su política sanitaria, su climatología, su suelo y sus aguas, sus condiciones sanitarias, su demografía, su morbicidad y su mortalidad. 2 volúmenes (Primero y Segundo) (Edición preparada por Carmen del Moral)* (Madrid: Editora Nacional, 1979).

Heredia y Larrea, Publio, *El testamento fonográfico* (Madrid: La Revista Política, 1895).

Hijos de Blas Cuesta, *Hijos de Blas Cuesta* (catalogue) (Valencia, ca. 1900).

Hijos de Blas Cuesta, *Hijos de Blas Cuesta* (advertisement), *Boletín fonográfico*, 27 (1901), p. 42.

Hugens, Armando, *Hugens y Acosta* (advertisement), *La correspondencia de España*, 2 January 1900, p. 3.

Hugens, Armando, *Sociedad Fonográfica Española* (advertisement), *Guía de la Coronación hecha expresamente para los forasteros que visiten Madrid en las fiestas que se celebrarán durante el mes de Mayo de 1902 con motivo de la coronación de S.M. el Rey D. Alfonso XIII*, ed. Gerardo Pardo (Madrid: publishing house unknown, 1902), pp. 96–7.

Indestructible Records, *Indestructible Records* (catalogue) (unknown location, 1908).

Johnson, Edward H., "A Wonderful Invention. Speech Capable of Infinite Repetition from Automatic Records", *Scientific American*, 17 November 1877, p. 304.

K. [name unknown], "Crónicas madrileñas", *La correspondencia de España*, 12 March 1897, p. 1.

Kardec, Allan, *Espiritismo experimental: El libro de los médiums. Guía de los médiums y de las evocaciones* (Barcelona: Carbonell y Esteva, 1904).

Kelly, Alan, *The Gramophone Company Limited, The Spanish Catalogue: Including Portuguese Recordings* (London: self-published, 2006).

La fonográfica madrileña, *La fonográfica madrileña* (advertisement), *ABC*, 2 April 1903, p. 4.

La fonográfica madrileña, *La fonográfica madrileña* (advertisement), *El heraldo de Madrid*, 6 May 1902, p. 4.

La fonográfica madrileña, *La fonográfica madrileña* (advertisement), *Nuevo Mundo*, 18 March 1903, p. 2.

La fonográfica madrileña, *La fonográfica madrileña* (advertisement), *ABC*, 28 January 1905.

Lioret, Henri, *Description et Prix des Differents Modeles de Lioretgraph* (catalogue) (Paris, ca. 1900).

López Alué, [first name unknown], "De La Granja", *El imparcial*, 18 September 1882, p. 2.

M. [full name unknown], "Crónicas madrileñas", *La época*, 22 December 1904, p. 1.

Marín, A.[first name unknown], "La impresión de fonogramas", *Boletín fonográfico*, 2 (1900), pp. 19–22.

Marqués de Alta-Villa, "Fonografía. Cuestión palpitante", *El cardo*, 8 April 1901, pp. 14–15.

Mathieu-Villon, A. [first name unknown], *El fonógrafo* (Madrid: A. Avial, 1893).

Mhartín y Guix, Enrique, *Reglamentación oficial de sociedades, reuniones y manifestaciones públicas* (Barcelona: José Cunill y Sala, 1902).

Moreno Casas, Manuel, *Centro fonográfico comercial* (advertisement), *La vanguardia*, 15 November 1900, p. 8.

Neuvillette, Christián de, "Crónicas de sociedad", *La correspondencia de España*, 26 August 1901, p. 2.

Odeón and Fonotipia, *Discos impresionados por las dos caras. Marcas Odeón y Fonotipia* (Catalogue) (Barcelona, undated).

Odeón and Fonotipia, *Catálogo General de Discos Odeón y Fonotipia* (Catalogue) (Barcelona, undated).

Pando y Valle, Jesús, *Regeneración económica. Croquis de un libro para el pueblo* (Madrid: Imprenta de Ricardo Rojas, 1897).

Pathé Freres London Limited, *Pathé Frères London Limited* (catalogue) (London, 1900).

Pernat, O. [first name unknown], *O. Pernot* [sic] (advertisement), *La vanguardia*, 13 January 1901, p. 3.

Pitis, [first name unknown], "Un charro filósofo", *El adelanto de Salamanca*, 24 August 1903, p. 1.

Questel – Edison Workshop in Paris, *Fonógrafo* (advertisement), *El imparcial*, 2 July 1895, p. 4.

Retana, Wenceslao, "Unos días en Huesca (Impresiones de un ex gobernador)", *Por esos mundos*, 155 (1907), pp. 514–26.

Roselló, [first name unknown], *Óptico Roselló* (advertisement), *La vanguardia*, 18 April 1898, p. 3.

Roselló, [first name unknown], *Óptico Roselló* (advertisement), *La vanguardia*, 7 May 1899, p. 1.

Roselló, [first name unknown], *Óptico Roselló* (advertisement), *Las noticias*, 7 May 1899, p. 2.

Roselló, [first name unknown], *Óptico Roselló* (advertisement), *La vanguardia*, 7 May 1899, p. 3.

Rosillo, Juan, *Sociedad artístico-fonográfica* (advertisement), *La publicidad*, 6 June 1899, p. 3.

Royo Villanova, Antonio, *La regeneración. El problema político* (Madrid: Hijos de M.G. Hernández, 1899).

Simiand, François, "Le salaire des ouvriers des mines de charbon en France", *Journal de la société statistique de Paris*, 49 (1908), pp. 13–29.

Sociedad Anónima Fonográfica, *Los reyes – Ningún regalo tan bonito y barato para este día como un fonógrafo, con 6 cilindros: 80 pesetas* (advertisement), *La correspondencia de España*, 3 January 1901.

Sociedad El Fonógrafo, *Sesiones de fonógrafo* (advertisement), *La correspondencia de España*, 31 January 1879, p. 1.

Sociedad El Fonógrafo, *Sesiones de fonógrafo* (advertisement), *El campo*, 16 February 1879, p. 16.

Sociedad Fonográfica Española Hugens y Acosta, *Catálogo* (Madrid, 1900).

Società Italiana di Fonotipia, *Catálogo* (Milan, 1907).

Solorno, Pascual del, "El fonógrafo", *Revista del Ateneo Escolar de Guadalajara*, 5 August 1882, pp. 53–5.

Sorrarain, R [first name unknown] de, "La ciencia actual", *La ilustración española e hispanoamericana*, 1 August 1888, p. 3.

Barón de Stoff, "Noticias telegráficas", *El globo*, 3 February 1897, p. 2.

T. [full name known], "La señorita Concha Sanz Arnal", *Boletín fonográfico*, 21 (1900), pp. 344–6.

T. [full name unknown], "El gramófono", *Las provincias*, 22 March 1904, p. 2.

T[altavull], A [first name unknown], "Fonógrafos automáticos", *Boletín fonográfico*, 14 (1900), pp. 216–17.

T[altavull], A [first name unknown], "Instrucciones especiales para el manejo del fonógrafo", *Boletín fonográfico*, 22 (1900), pp. 362–3.

T[altavull], A [first name unknown], "El cilindro fonográfico", *Boletín fonográfico*, 30 (1901), pp. 101–2.

T[altavull], A [first name unknown], "Impresión de cilindros", *Boletín fonográfico*, 33 (1901), p. 150.

Teatro Circo Balear, *Teatro Circo Balear* (advertisement), *Las Baleares*, 16 March 1895, p. 3.

Torres Orive, Manuel, "El fonógrafo", *Las provincias*, 30 June 1899, p. 2.

Un diletante [pseudonym], "Música clásica. Cuartetos en el Conservatorio", *La correspondencia de España*, 16 December 1880, pp. 2 and 4.

Un patriota [pseudonym], "Fonograma", *El cardo*, 28 February 1901, p. 15.

Uña y Sarthou, Juan, *Las asociaciones obreras en España* (Madrid: G. Juste, 1900).

Ureña, Álvaro, "Comunicado", *La correspondencia militar*, 16 February 1900, p. 2.

Ureña, Álvaro, [untitled], *El cardo*, 8 March 1901, p. 15.

Ureña, Álvaro, *Álvaro Ureña* (advertisement), *La época*, 15 December 1902, p. 3.

Ureña, Álvaro, *Álvaro Ureña* (advertisement), *ABC*, 16 April 1903, p. 4.

Ureña, Álvaro, *Álvaro Ureña* (advertisement), *Diario de Córdoba*, 7 May 1904, p. 4.

Ureña, Álvaro, *Álvaro Ureña* (advertisement), *El adelanto de Salamanca*, 31 July 1904, p. 4.

Valmont, Blanca, "Crónica", *La última moda*, 269 (1893), 1–2.

Vargas, Martín de, "Julián Biel", *El heraldo de Madrid*, 21 September 1899, p. 3.

Verdú, Josep, "La cansó del fonógrafo", *La Renaxensa*, 30 October 1879, pp. 38–40.

Vicente, Eusebio, *Compañía de máquinas parlantes* (advertisement), *El pueblo*, 1 July 1907, no page number.

Victor, *His Masters' Voice. Celebrity Records by International Artists. With Special Biographies and Photographs* (Catalogue) (no location, 1914).

Victor, *El libro Victrola de la ópera* (Camden, NJ: Victor Talking Machine Company, 1925).

Vinaixa, [first name unknown], "Noticias", *El liberal*, 19 October 1899, p. 2.

Viuda de Aramburo, *Viuda de Aramburo* (advertisement), *ABC*, 22 May 1897, p. 3.

Willmann, Carl, *La magia moderna de salón* (Valencia: Pascual Eguilar, 1897).

P. T. V. [full name unknown], "El fonógrafo", *Teléfono catalán*, 25 May 1879, p. 1.

X. [full name unknown], "Sección varia. El fonógrafo", *La antorcha*, 5 February 1883, p. 3.

Secondary sources

Adorno, Theodor W., "The Curves of the Needle", trans. Thomas Y. Levin, *October*, 55 (1990), pp. 48–55.

Andrés-Gallego, José, *Un 98 distinto. Regeneración, desastre, regeneracionismo* (Madrid: Ediciones Encuentro, 1998).

Ashby, Arved, *Absolute Music, Mechanical Reproduction* (Oakland: University of California Press, 2010).

Baratas Díaz, Alfredo, "La ciencia española ante la crisis del 98: semillas, frutos y agostamiento", *Cuadernos de Historia Contemporánea*, 20 (1998), pp. 151–63.

Barreiro, Javier, and Gabriel Marro, *Primeras grabaciones fonográficas en Aragón. 1898-1903. Una colección de cilindros de cera* (Zaragoza: Coda OUT/Gobierno de Aragón, 2007).

Barreiro, Javier, *Antiguas grabaciones fonográficas aragonesas, 1898-1907: la colección de cilindros para fonógrafo de Leandro Pérez* (Zaragoza: Coda OUT/Gobierno de Aragón, 2010).

Belchior, Susana, "Manufacturing Records", *The Lindstrom Project*, vol. 9, ed. Pekka Gronow, Christiane Hofer, and Mathias Böhm (Vienna: Gesellschaft für Historische Tonträger, 2017).

Biblioteca Nacional de Catalunya, *Cilindres de cera de la col·lecció Mariano Gómez Montejano* [non-commercial CD], Biblioteca Nacional de Catalunya, CD 4175.

Biddle, Ian, "Madrid's Great Sonic Transformation: Sound, Noise, and the Auditory Commons of the City in the Nineteenth Century", *Journal of Spanish Cultural Studies*, 19:3 (2019), pp. 227–40.

Blake, Andrew, "Recording Practices and the Role of the Producer", *The Cambridge Companion to Recorded Music*, ed. Nicholas Cook, Eric Clarke, Daniel Leech-Wilkinson, and John Rink (Cambridge: Cambridge University Press, 2009), pp. 36–53.

Bolig, John R., *The Victor Red Seal Discography Vol. II. Double-Sided Series to 1930* (Denver: Mainspring Press, 2004).

Bolig, John R., *The Victor Discography. Green, Blue, and Purple Labels (1910-1926)* (Denver: Mainspring Press, 2006).

Bowers, Nathan David, "Creating a Home Culture for the Phonograph: Women and the Rise of Sound Recordings in the United States, 1877–1913" (Ph.D. diss., University of Pittsburgh, 2007).

Camarasa, Josep M., and Antoni Roca Rosell, eds., *Ciència i tècnica als Països Catalans: una aproximació biogràfica* (Barcelona: Fundació Catalana per a la Recerca, 1995).

Cañardo, Marina, *Fábricas de músicas. Comienzos de la industria discográfica en Argentina (1919-1930)* (Buenos Aires: Gourmet Ediciones, 2017).

Cañellas, Cèlia, and Rosa Torán, "Heterogeneïtat urbana, desplaçaments geogràfics i canvis polítics", *La formació de l'Eixample de Barcelona: aproximacions a un fenomen urbà*, ed. Santi Barjau et al. (Barcelona: Olimpíada Cultural, 1990), pp. 189–202.

Carabell, Paula, "Photography, Phonography and the Lost Object", *Perspectives of New Music*, 40:1 (2002), pp. 176–89.

Carrasco Mas, Sara, *Puesta en valor de la colección de cilindros de cera del Museo de Historia de la Telecomunicación Vicente Miralles Segarra* (undergraduate dissertation, Universitat Politècnica de València, 2018).

Casado de Otaola, Santos, *Naturaleza patria: ciencia y sentimiento de la naturaleza en la España del Regeneracionismo* (Madrid: Marcial Pons, 2010).

Casares, Emilio, "Leocadia Alba", *Diccionario de la zarzuela. España e Hispanoamérica*, ed. Emilio Casares (Madrid: ICCMU, 2008), 1:27.

Casares, Emilio, "Compañías", *Diccionario de la zarzuela. España e Hispanoamérica*, ed. Emilio Casares (Madrid: ICCMU, 2008), 1:525–39.

Casares, Emilio, "Voz", *Diccionario de la zarzuela. España e Hispanoamérica*, ed. Emilio Casares (Madrid: ICCMU, 2008), 2:941–44.

Casassas, Jordi, *La voluntat i la quimera. El noucentisme català entre la Renaixença i el marxisme* (Barcelona: Pòrtic, 2017).

Cayuela Fernández, José Gregorio, "Proyectos de sociedad y nación: la crisis del concepto de España en el 98", *Sociabilidad fin de siglo. Espacios asociativos en torno a 1898*, ed. Isidro Sánchez Sánchez and Rafael Villena Espinosa (Cuenca: Ediciones de la Universidad de Castilla-La Mancha, 1999), pp. 45–72.

Centro Andaluz del Flamenco, *Cilindros de cera, fondos del Centro Andaluz de Flamenco: primeras grabaciones del flamenco* (Mairena de Aljarafe: Calé Records, 2003).

Chacón Delgado, Pedro José, *Historia y nación. Costa y el regeneracionismo en el fin de siglo* (Santander: Ediciones de la Universidad de Cantabria, 2013).

Chamoux, Henri, "La diffusion de l'enregistrement sonore en France à la Belle Époque (1893–1914), artistes, industriels et auditeurs du cylindre et du disque" (Ph.D. diss., Université de Paris I, 2015).

Chanan, Michael, *Repeated Takes: A Short History of Recording and Its Effects on Music* (New York: Verso, 1995).

Charbon, Paul, *La machine parlante* (Paris: J.P.Gyss, 1981).

Chew, V. K., *Talking Machines: 1877–1914: Some Aspects of the Early History of the Gramophone* (London: Her Majesty's Stationery Office, 1967).

Cisneros Solá, María Dolores, "La obra para voz y piano de Manuel de Falla: Contexto artístico-cultural, proceso creativo y primera recepción" (Ph.D. diss., Universidad Complutense de Madrid, 2018).

Climent, José, "Lamberto Alonso", *Diccionario de la música valenciana*, ed. Emilio Casares, Vicente Galbis López, and Rafael Díaz Gómez (Valencia: Institut Valencià de la Músical, 2006), 1:41–2.

Cook, Nicholas, "Performance Analysis and Chopin's Mazurkas", *Musicae Scientiae*, 11 (2007), pp. 183–207.

Cook, Nicholas, "The Ghost in the Machine: Towards a Musicology of Recordings", *Musicae Scientiae*, 14 (2010), pp. 3–21.

Cook, Nicholas, *Beyond the Score: Music as Performance* (New York: Oxford University Press, 2013).

Corominas, Miquel, "Les societats de l'eixample", *La formació de l'Eixample de Barcelona: aproximacions a un fenomen urbà*, ed. Santi Barjau et al. (Barcelona: Olimpíada Cultural, 1990), pp. 43–59.

Cueva Merino, Julio de la, "Clericalismo y movilización católica durante la restauración", *Clericalismo y asociacionismo católico en España: de la restauración a la transición: un siglo entre el palio y el consiliario*, ed. Julio de la Cueva Merino and Ángel Luis López Villaverde (Cuenca: Ediciones de la Universidad de Castilla la Mancha, 2005), pp. 27–50.

Day, Timothy, *A Century of Recorded Music: Listening to Musical History* (New Heaven, CT, and London: Yale University Press, 2002).

Deleito y Piñuela, José, *Origen y apogeo del "género chico"* (Madrid: Revista de Occidente, 1949).

Denning, Michael, *Noise Uprising. The Audiopolitics of a World Musical Revolution* (New York: Verso, 2015).

Díaz Frene, Jaddiel, "A las palabras ya no se las lleva el viento: apuntes para una historia cultural del fonógrafo en México (1876–1924)", *Historia Mexicana*, 66 (2016), pp. 257–98.

Doughy, Karolina, and Maja Lagerkvist, "The Pan Flute Musicians at Sergels Torg: Between Global Flows and Specificities of Place", *Researching and Representing Mobilities. Transdisciplinary Encounters*, ed. Lesley Murray and Sara Upstone (London: Palgrave MacMillan, 2014), pp. 150–69.

Elena, Alberto, and Javier Ordóñez, "Science, Technology, and the Spanish Colonial Experience in the Nineteenth Century", *Osiris*, 15 (2000), pp. 70–82.

Feaster, Patrick, "Framing the Mechanical Voice: Generic Conventions of Early Sound Recording", *Folklore Forum*, 32 (2001), pp. 57–102.

Feaster, Patrick, "'The Following Record': Making Sense of Phonographic Performance, 1877–1908" (Ph.D. diss., Indiana University, 2007).

Feaster, Patrick, "'Rise and Obey the Command': Performative Fidelity and the Exercise of Phonographic Power", *Journal of Popular Music Studies*, 24 (2012), pp. 357–95.

Fernández, Luis Miguel, *Tecnología, espectáculo, literatura: dispositivos ópticos en las letras españolas de los siglos XVIII y XIX* (Santiago de Compostela: Publicacións Universidade de Santiago de Compostela, 2006).

Fernández Paradas, Mercedes, "La industria eléctrica y su actividad en el negocio del alumbrado en España (1901–1935)", *Ayer*, 71 (2008), pp. 245–65.

Fernández Roca, Francisco Javier, "El salario industrial en Sevilla: 1900–1975", *Industria y clases trabajadoras en la Sevilla del siglo XX*, ed. Carlos Arenas (Sevilla: Secretariado de Publicaciones de la Universidad de Sevilla, 1994), pp. 115–42.

Ferrater Mora, José, *Three Spanish Philosophers* (Albany: State University of New York Press, 2003).

Gabriel, Pere, "Red Barcelona in the Europe of War and Revolution, 1914–1930", *Red Barcelona. Social Protest and Labour Mobilization in the 20th Century*, ed. Angel Smith (London: Routledge, 2002), pp. 44–65.

Garcia Espuche, Albert, *El Quadrat d'Or: centro de la Barcelona modernista: la formación de un espacio urbano privilegiado* (Barcelona: Ajuntament de Barcelona, Lunwerg Editores, 2002).

Garcia Espuche, Albert, "El centre residencial burgès (1860–1914)", *La formació de l'Eixample de Barcelona: aproximacions a un fenomen urbà*, ed. Santi Barjau et al. (Barcelona: Olimpíada Cultural, 1990), pp. 203–21.

Gauß, Stefan, "Listening to the Horn: On the Cultural History of the Phonograph and the Gramophone", *Sounds of Modern History: Auditory Cultures in 19th- and 20th-Century Europe*, ed. Daniel Morat (Oxford, New York: Berghahn, 2014), pp. 71–100.

Gelatt, Roland, *Edison's Fabulous Phonograph 1877–1977* (London: MacMillan, 1977).

Girón Sierra, Álvaro, and Jorge Molero-Mesa, "The Rose of Fire. Anarchist Culture, Urban Spaces and Management of Scientific Knowledge in a Divided City", *Barcelona: An Urban History of Science and Modernity, 1888–1929*, ed. Oliver Hochadel and Agustí Nieto-Catalán (Oxford and New York: Routledge, 2006), pp. 115–35.

Gitelman, Lisa, *Scripts, Grooves, and Writing Machines. Representing Technology in the Edison Era* (Stanford: Stanford University Press, 1999).

Gitelman, Lisa, "Reading Music, Reading Records, Reading Race: Musical Copyright and the U.S. Copyright Act of 1909", *The Musical Quarterly*, 81:2 (1997), pp. 265–90.

Gómez-Montejano, Mariano, "El fonógrafo en España (segunda parte)", *Girant a 78 rpm*, 6 (2004), pp. 9–11.

Gómez-Montejano, Mariano, "El fonógrafo en España (tercera parte)", *Girant a 78 rpm*, 8 (2005), pp. 6–8.

Gómez-Montejano, Mariano, *El fonógrafo en España. Cilindros españoles* (Madrid: self-published, 2005).

González Peña, María Luz, "Rafael Bezares", *Diccionario de la zarzuela. España e Hispanoamérica*, ed. Emilio Casares (Madrid: ICCMU, 2008), 1:261.

Gopinath, Sumanth, and Jason Stanyek, "Anytime, Anywhere? An Introduction to the Devices, Markets, and Theories of Mobile Music", *The Oxford Handbook of Mobile*

Music Studies, ed. Sumanth Gopinath and Jason Stanyek (New York: Oxford University Press, 2014), pp. 1–34.

Gronow, Pekka, "The Record Industry: The Growth of a Mass Medium", *Popular Music*, 3 (1983), pp. 53–75.

Gronow, Pekka, and Ilpo Saunio, *An International History of the Recording Industry* (London and New York: Cassell, 1999).

Gronow, Pekka, and Björn Englund, "Inventing Recorded Music: The Recorded Repertoire in Scandinavia, 1899–1925", *Popular Music*, 26:2 (2007), pp. 281–304.

Guereña, Jean-Louis, "La sociabilidad en la España contemporánea", *Sociabilidad fin de siglo. Espacios asociativos en torno a 1898*, ed. Isidro Sánchez Sánchez and Rafael Villena Espinosa (Cuenca: Ediciones de la Universidad de Castilla-La Mancha, 1999), pp. 15–44.

Harrison, Joseph, "Tackling National Decadence: Economic Regenerationism in Spain after the Colonial Debacle", *Spain's 1898 Crisis. Regenerationism, Modernism, Post-Colonialism*, ed. Joseph Harrison and Alan Hoyle (Manchester and New York: Manchester University Press, 2000), pp. 55–67.

Hibbs, Solange, and Caroline Fillière, "Introducción", *Los discursos de la ciencia y de la literatura en España (1875–1906)*, ed. Solange Hibbs and Caroline Fillière (Madrid: Academia del Hispanismo, 2015), pp. 13–28.

Hochadel, Oliver, and Agustí Nieto-Catalán, "Introduction", *Barcelona: An Urban History of Science and Modernity, 1888–1929*, ed. Oliver Hochadel and Agustí Nieto-Catalán (Oxford and New York: Routledge, 2006), pp. 1–21.

Hochadel, Oliver, and Laura Valls, "Civic Nature. The Transformation of the Parc de la Ciutadella into a Space for Popular Science", *Barcelona: An Urban History of Science and Modernity, 1888–1929*, ed. Oliver Hochadel and Agustí Nieto-Catalán (Oxford and New York: Routledge, 2016), pp. 25–45.

Hogg, Bennett, "The Cultural Imagination of the Phonographic Voice, 1877–1940" (Ph.D. diss., University of Newcastle, 2008).

Kahn, Douglas, "Introduction", *Wireless Imagination. Sound, Radio, and the Avant-Garde*, ed. Douglas Kahn and Gregory Whitehead (Cambridge, MA, and London: The MIT press, 1992), pp. 1–30.

Katz, Mark, *Capturing Sound: How Technology Has Changed Music* (Berkeley: University of California Press, 2010).

Katz, Mark, "The Amateur in the Age of Mechanical Music", *The Oxford Handbook of Sound Studies*, ed. Trevor Pinch and Karin Bijsterveld (New York: Oxford University Press, 2011), pp. 459–78.

Kenney, William Howland, *Recorded Music in American Life: The Phonograph and Popular Memory, 1890–1945* (New York and Oxford: Oxford University Press, 1999).

Kinnear, Michael, *The Gramophone Company's First Indian Recordings 1899–1909* (Bombay: Popular Prakashan, 1994).

Knight, David, "La popularización de la ciencia en la Inglaterra del siglo XIX", *La ciencia y su público*, ed. Javier Ordóñez y Alberto Elena (Madrid: CSIC, 1990), pp. 311–30.

Kraft, James P., *Stage to Studio. Musicians and the Sound Revolution, 1890–1950* (Baltimore and London: The Johns Hopkins University Press, 1996).

Kursell, Julia, "A Gray Box: The Phonograph in Laboratory Experiments and Fieldwork, 1900–1920", *The Oxford Handbook of Sound Studies*, ed. Trevor Pinch and Karin Bijsterveld (New York: Oxford University Press, 2011), pp. 176–94.

Landaberea, Jaione, "Ybarra Family Fonds", *Eresbil. Archivo de la Música Vasca*, http://www.eresbil.com/web/ybarra/Pagina.aspx?moduleID=2717&lang=en.

Leech-Wilkinson, Daniel, "Portamento and Musical Meaning", *Journal of Musicological Research*, 25 (2006), pp. 233–61.

Leech-Wilkinson, Daniel, "Sound and Meaning in Recordings of Schubert's 'Die junge Nonne'", *Musicae Scientiae*, 11 (2007), pp. 209–36.

Leech-Wilkinson, Daniel, *The Changing Sound of Music* (London: CHARM, 2009).

Levin, Thomas Y., "For the Record: Adorno on Music in the Age of Its Technological Reproducibility", *October*, 55 (1990), pp. 23–47.

Litvak, Lily, *El tiempo de los trenes: el paisaje español en el arte y la literatura del realismo: (1849–1918)* (Madrid: Ediciones del Serbal, 1991).

Llano, Samuel, *Discordant Notes. Marginality and Social Control in Madrid, 1850–1930* (New York: Oxford University Press, 2018).

Llopart, Maria Dolors, "Les cases de l'Eixample de portes endins", *La formació de l'Eixample de Barcelona: aproximacions a un fenomen urbà*, ed. Santi Barjau (Barcelona: Olimpíada Cultural, 1990), pp. 115–27.

López Piñero, José María, ed., *La ciencia en la España del siglo XIX* (Madrid: Marcial Pons, 1992).

Losa, Leonor, *Machinas fallantes: a música gravada em Portugal no início do século XX* (Lisboa: Tinta da China, 2013).

Lotz, Rainer E., "On the History of Lindström AG", *The Lindström Project*, ed. Pekka Gronow and Christiane Hofer (Vienna: Gesellschaft für Historische Tonträger, 2009), 1:11–22.

Lysloff, René T. A., and Leslie C. Gay, Jr., eds., *Music and Technoculture* (Middletown, CT: Wesleyan University Press, 2003).

Maisonneuve, Sophie, "De la 'machine parlante' à l'auditeur: le disque et la naissance d'une nouvelle culture musicale dans les années 1920–1930", *Terrain*, 37 (2001), pp. 11–28.

Maisonneuve, Sophie, "De la machine parlante au disque. Une innovation technique, commerciale et culturelle", *Vingtième siècle. Revue d'histoire*, 92 (2006), pp. 17–31.

Mallada, Lucas, *La futura revolución española y otros escritos regeneracionistas*, ed. Francisco J. Ayala-Carcedo and Steven L. Driever (Madrid: Biblioteca Nueva, 1998).

Martín Ballester, Carlos, "Frequently Asked Questions", http://www.carlosmb.com/pruebas/e_preguntas.php, accessed December 2018.

Martín de Sagarmínaga, Joaquín, *Diccionario de cantantes líricos españoles* (Madrid: Fundación Caja de Madrid, 1997).

Martínez Roda, Federico, *Valencia y las Valencias: su historia contemporánea (1800–1975)* (Valencia: Fundación Universitaria San Pablo CEU, 1998).

Martland, Peter, *Recording History. The British Record Industry, 1888–1931* (Lanham, MD, Toronto, and Plymouth, UK: The Scarecrow Press, 2013).

Martorell, Miguel, and Santos Juliá, *Manual de Historia Política y Social de España (1808–2018)* (Barcelona: RBA Libros, 2019).

Marty, Daniel, *Histoire illustré du phonographe* (Paris: Lazarus, 1979).

Moral Ruiz, Carmen del, *El género chico* (Madrid: Alianza, 2004).

Moreda Rodríguez, Eva, "Travelling Phonographs in *fin de siècle* Spain: Recording Technologies and National Regeneration in Ruperto Chapí's *El fonógrafo ambulante*", *Journal of Spanish Cultural Studies*, 20:3 (2019), pp. 241–55.

Moreda Rodríguez, Eva, "Recording *zarzuela grande* in Spain in the Early Days of the Phonograph and Gramophone", *Music, Nation and Region in the Iberian Peninsula*, ed. Samuel Llano, Matthew Machin-Autenrieth, and Salwa Castelo-Branco (Champaign: University of Illinois Press, forthcoming in 2021).

Moten, Fred, "The Phonographic *mise-en-scène*", *Cambridge Opera Journal*, 16 (2004), pp. 269–81.

Muñoz, Matilde, *Historia de la zarzuela y el género chico* (Madrid: Tesoro, 1946).

Muñoz Duato, Jacobo, "Vicente Gómez Novella, 1871–1956" (Ph.D. diss., University of Valencia, 2015).

Naranjo Santana, Mari Carmen, *Cultura, ciencia y sociabilidad en Las Palmas de Gran Canaria en el siglo XIX: el Gabinete Literario y el Museo Canario* (Rivas Vaciamadrid: Mercurio, 2016).

Ospina-Romero, Sergio, "Ghosts in the Machine and Other Tales around a 'Marvelous Invention': Player Pianos in Latin America in the Early Twentieth Century", *Journal of the American Musicological Society*, 72:1 (2019), pp. 1–42.

Oyón, José Luis, *La quiebra de la ciudad popular: espacio urbano, inmigración y anarquismo en la Barcelona de entreguerras, 1914–1936* (Barcelona: Ediciones del Serbal, 2008).

Patmore, David, "Selling Sounds: Recording and the Record Business", *The Cambridge Companion to Recorded Music*, ed. Nicholas Cook, Eric Clarke, Daniel Leech-Wilkinson, and John Rink (Cambridge: Cambridge University Press, 2009), pp. 120–39.

Pérez Castroviejo, Pedro María, "Precios, salarios reales y estaturas en el curso de la industrialización del País Vasco, 1880–1936", *Salud y ciudades en España, 1880–1940. Condiciones ambientales, niveles de vida e intervenciones sanitarias*, ed. Francisco Muñoz, Pedro Fatjó, and Josep Pujol-Andreu (Barcelona: Universitat Autònoma de Bacelona), p. 8.

Perlman, Marc, "Golden Ears and Meter Readers: The Contest for Epistemic Authority in Audiophilia", *Social Studies of Science*, 34:5 (2004), pp. 783–807.

Perramon i Basqué, Josep, "Presentació de l'Aspe, Associació per a la Salvaguarda del Patrimoni Enregistrar", *Girant a 78 rpm*, 1:1 (2002), pp. 3–4.

Perrone, Raffaella, "Espacio teatral y escenario urbano. Barcelona entre 1840 y 1923" (Ph.D. diss., Universitat Politècnica de Catalunya, 2011).

Philip, Robert, *Performing Music in the Age of Recording* (New Haven, CT: Yale University Press, 2004).

Picker, John M., "The Victorian Aura of the Recorded Voice", *New Literary History*, 32 (2001), pp. 769–86.

Picker, John M., *Victorian Soundscapes* (New York: Oxford University Press, 2003).

Pinch, Trevor, and Karin Bijsterveld, "New Keys to the World of Sound", *The Oxford Handbook of Sound Studies*, ed. Trevor Pinch and Karin Bijsterveld (New York: Oxford University Press, 2011), pp. 3–36.

Plack, Rebecca, "The Substance of Style: How Singing Creates Sound in Lieder Recordings, 1902–1939" (Ph.D. diss., Cornell University, 2008).

Puerto Sarmiento, Francisco Javier, "Ciencia y farmacia en la España decimonónica", *La ciencia en la España del siglo XIX*, ed. José María López Piñero (Madrid: Marcial Pons, 1992), pp. 153–92.

Read, Oliver, and Walter L. Welch, *From Tin Foil to Stereo Evolution of the Phonograph* (Indiana: Howard W. Sams, 1976).

Regidor Arribas, Ramón, *La voz en la zarzuela* (Madrid: Real Musical, 1991).

Resina, Joan Ramon, *La vocació de modernitat de Barcelona: auge i declivi d'una imatge urbana* (Barcelona: Galàxia Gutenberg, 2008).

Rider, Robin E., "El experimento como espectáculo", *La ciencia y su público*, ed. Javier Ordóñez y Alberto Elena (Madrid: CSIC, 1990), pp. 113–45.

Riego Amezaga, Bernardo, "Imágenes fotográficas y estrategias de opinión pública: los viajes de la Reina Isabel II por España (1858–1866)", *Patrimonio Nacional*, 139 (1999), pp. 2–13.

Riego Amezaga, Bernardo, *La construcción social de la realidad a través de la fotografía y el grabado informativo en la España del siglo XIX* (Santander: Universidad de Cantabria, 2001).

Riego Amezaga, Bernardo, "Visibilidades diferenciadas: usos sociales de las imágenes en la España isabelina", *Ojos que ven, ojos que leen. Textos e imágenes en la España Isabelina*, ed. Marie-Linda Ortega (Madrid: Visor/ Presses Universitaires de Marne-La-Vallée, 2004), pp. 57–66.

Riego Amezaga, Bernardo, "La representación en las imágenes fotográficas: discursos y divergencias de cada tiempo histórico", *Fotografía e Historia. III Encuentro en Castilla la Mancha*, ed. Irma Fuencisla Álvarez Delgado and Ángel Luis López Villaverde (Cuenca: UCLM, 2009), pp. 67–81.

Riera i Tuèbols, Santiago, *Història de la ciència a la Catalunya moderna* (Lleida i Vic: Pagès/Eumo, 2003).

Roca Rosell, Antoni, "El discurso civil en torno a la ciencia y la técnica", *El regeneracionismo en España. Política, educación, ciencia y sociedad*, ed. Vicente Salavert and Manuel Suárez Cortina (Valencia: Universitat de Valencia, 2007), pp. 241–59.

Rodríguez Gutiérrez, Borja, "Ciencia e imágenes de la ciencia", *Los discursos de la ciencia y de la literatura en España (1875–1906)*, ed. Solange Hibbs y Caroline Fillière (Madrid: Academia del Hispanismo, 2015), pp. 225–46.

Rothenbuhler, Eric W., and John Durham Peters, "Defining Phonography: An Experiment in Theory", *The Musical Quarterly*, 81:2 (1997), pp. 242–64.

Roy, Élodie A., "'You Ought to See My Phonograph': The Visual Wonder of Recorded Sound", *The Making of English Popular Culture*, ed. John Storey (Oxford and New York: Routledge, 2016), pp. 184–97.

Ruiz-Castell, Pedro, "Scientific Instruments for Education in Early Twentieth-Century Spain", *Annals of Science*, 65:4 (2008), pp. 519–27.

Sablonnière, Catherine, "Literatura científica y vulgarización de las ciencias ante la crisis de la ciencia en la España finisecular: ¿hacia una poética del discurso científico?", *Los discursos de la ciencia y de la literatura en España (1875–1906)*, ed. Solange Hibbs and Caroline Fillière (Madrid: Academia del Hispanismo, 2015), pp. 305–18 (pp. 313–14).

Sagarra, Ferran, "Barcelona dins del projecte industrialista català", *La formació de l'Eixample de Barcelona: aproximacions a un fenomen urbà*, ed. Santi Barjau et al. (Barcelona: Olimpíada Cultural, 1990), pp. 13–25.

Salavert, Vicente, and Suárez Cortina, Manuel, "Introducción", *El regeneracionismo en España. Política, educación, ciencia y sociedad*, ed. Vicente Salavert and Manuel Suárez Cortina (Valencia: Universitat de Valencia, 2007), pp. 9–19.

Sánchez Rebanal, Fernando, "La vida escénica en la ciudad de Santander entre 1895 y 1904" (Ph.D. diss., Universidad Nacional de Educación a Distancia, 2014).

Sánchez Ron, José Manuel, "Investigación científica, desarrollo tecnológico y educación en España (1900–1950)", *Arbor*, 96 (1992), pp. 33–74.

Sánchez Sánchez, Isidro, "Las luces del 98. Sociedades eléctricas en la España finisecular", *Sociabilidad fin de siglo. Espacios asociativos en torno a 1898*, ed. Isidro Sánchez Sánchez and Rafael Villena Espinosa (Cuenca: Ediciones de la Universidad de Castilla-La Mancha, 1999), pp. 151–223.

Sastre-Juan, Jaume, and Jaume Valentines-Álvarez, "Technological Fun. The Politics and Geographies of Amusement Parks", *Barcelona: An Urban History of Science and Modernity, 1888–1929*, ed. Oliver Hochadel and Agustí Nieto-Catalán (Oxford and New York: Routledge, 2016), pp. 115–35.

Schmidt-Horning, Susan, "Engineering the Performance: Recording Engineers, Tacit Knowledge, and the Art of Controlling Sound", *Social Studies of Science*, 34 (2004), pp. 703–31.

Seil, Michel, "Opera Singers on the Lindström Labels", *The Lindström Project*, ed. Pekka Gronow and Christiane Hofer (Vienna: Gesellschaft für Historische Tonträger, 2009), 1:41–2.

Serrano García, Rafael, *El Círculo de Recreo de Valladolid (1844–2010): ocio y sociabilidad en un espacio exclusivo* (Valladolid: Universidad de Valladolid, 2011).

Silva, João, "Mechanical Instruments and Phonography: The Recording Angel of Historiography", *Radical Musicology*, 6 (2012–2013), http://www.radical-musicology.org.uk/2012/DaSilva.htm, accessed June 2020.

Silva, João, *Entertaining Lisbon: Music, Theater, and Modern Life in the Late 19th Century* (New York: Oxford University Press, 2016).

Smith, Angel, "From Subordination to Contestation: The Rise of Labour in Barcelona, 1898–1918", *Red Barcelona. Social Protest and Labour Mobilization in the 20th Century*, ed. Angel Smith (London: Routledge, 2002), pp. 17–43.

Smith, Angel, "Barcelona through the European mirror: From Red and Black to Claret and Blue", *Red Barcelona. Social Protest and Labour Mobilization in the 20th Century*, ed. Angel Smith (London: Routledge, 2002), pp. 1–16.

Solà, Pere, *Els Ateneus obrers i la cultura popular a Catalunya (1900–1939): l'Ateneu Enciclopèdic Popular* (Barcelona: La Magrana, 1978).

Spottswood, Richard K., *Ethnic Music on Records: A Discography of Ethnic Recordings Produced in the United States, 1893 to 1942* (Urbana and Chicago: University of Illinois Press, 1990), vol. 1.

Stadler, Gustavus, "Never Heard Such a Thing. Lynching and Phonographic Modernity", *Social Text*, 28 (2010), pp. 87–110.

Sterne, Jonathan, *The Audible Past. Cultural Origins of Sound Reproduction* (Durham and London: Duke University Press, 2003).

Sutton, Allan, *Directory of American Disc Record Brands and Manufacturers, 1891–1943* (Westport and London: Greenwood Press, 1994).

Suárez Cortina, Manuel, "Sociedad, cultura y política en la España de entre siglos", *El regeneracionismo en España. Política, educación, ciencia y sociedad*, ed. Vicente Salavert and Manuel Suárez Cortina (Valencia: Universitat de Valencia, 2007), pp. 21–46.

Suisman, David, *Selling Sounds. The Commercial Revolution in American Music* (Cambridge, MA: Harvard University Press, 2009).

Suisman, David, "Sound, Knowledge, and the 'Immanence of Human Failure': Rethinking Musical Mechanization through the Phonograph, the Player-Piano, and the Piano", *Social Text*, 28 (2010), pp. 12–34.

Swyngedouw, Erik, *Liquid Power. Contested Hydro-Modernities in Twentieth-Century Spain* (Cambridge, MA: MIT Press, 2015).

Thompson, Emily, "Machines, Music, and the Quest for Fidelity: Marketing the Edison Phonograph in America, 1877–1925", *The Musical Quarterly*, 79:1 (1995), pp. 131–71.

Torrent i Marqués, Antoni, "Visió europea del naixement de l'enregistrament sonor", *Girant a 78 rpm*, 1:1 (2002), pp. 5–9.

Torrent i Marqués, Antoni, "Girant al voltant dels Berliner's", *Girant a 78 rpm*, 1:3 (2003), pp. 3–7.

Torrent i Marqués, Antoni, "Efectes secundaris dels discs a 78 rpm", *Girant a 78 rpm*, 2:4 (2004), pp. 5–12.

Torrent i Marqués, Antoni, "Una recerca interessant (1º part)", *Girant a 78 rpm*, 8 (2005), pp. 9–11.

Torrent i Marqués, Antoni, "Els primers enregistraments a casa nostra", *Girant a 78 rpm*, 11 (2008), pp. 6–13.

Torrent i Marqués, Antoni, "Sardanes a 78 rpm. Enregistraments durant els anys 1906–1907", *Girant a 78 rpm*, 15 (2009), pp. 13–17.

Torres y Oriol, Isidro, *Barcelona histórica antigua y moderna: guía general descriptiva e ilustrada* (Diputación y Ayuntamiento, 1905).

Uría, Jorge, "El nacimiento del ocio contemporáneo. Algunas reflexiones sobre el caso español", *Fiesta, juego y ocio en la historia*, ed. Ángel Vaca Lorenzo (Salamanca: Ediciones Universidad de Salamanca, 2002), pp. 347–82.

Velásquez Ospina, Juan Fernando, "(Re)sounding Cities: Urban Modernization, Listening, and Sounding Cultures in Colombia, 1886–1930" (Ph.D. diss., University of Pittsburgh, 2018).

Vest, Jacques, "Vox Machinae: Phonographs and the Birth of Sonic Modernity, 1877–1930" (Ph.D. diss., University of Michigan, 2018).

Villagrasa i Hernàndez, Fèlix, *Mancomunitat i ciència: la modernització de la cultura catalana* (Catarroja/Barcelona/Palma: Afers, 2015).

Wilson, Alexandra, "Galli-Curci Comes to Town", *The Arts of the Prima Donna in the Long Nineteenth Century*, ed. Rachel Cowgill and Hilary Poriss (New York, 2012).

Yasar, Kerim, *Electrified Voices. How the Telephone, Phonograph and Radio Shaped Modern Japan, 1898–1945* (New York and Chichester: Columbia University Press, 2018).

Ziegler, Susanne, *Die Wachszylinder des Berliner Phonogramm-Archivs* (Berlin: Ethnologisches Museum Staatliche Museen zu Berlin, 2005).

Zozaya, María, *Identidades en juego: formas de representación social del poder de la élite en un espacio de sociabilidad masculino, 1836–1936* (Madrid: Siglo XXI de España, 2015).

Zucconi, Benedetta, *Coscienza fonografica. La riflessione sul suono registrato nell'Italia del primo Novecento* (Napoli-Salerno: Orthotes, 2018).

Index